Hazardous Future

I0023450

Hazardous Future

Disaster, Representation and the Assessment of Risk

Edited by
Isabel Capeloa Gil and Christoph Wulf

DE GRUYTER

ISBN 978-3-11-055307-9
e-ISBN (PDF) 978-3-11-040661-0
e-ISBN (EPUB) 978-3-11-040680-1

Library of Congress Cataloging-in-Publication Data
A CIP catalog record for this book has been applied for at the Library of Congress.

Bibliographic information published by the Deutsche Nationalbibliothek
The Deutsche Nationalbibliothek lists this publication in the Deutsche Nationalbibliografie;
detailed bibliographic data are available on the Internet at http://dnb.dnb.de.

© 2017 Walter de Gruyter GmbH, Berlin/Munich/Boston
This volume is text- and page-identical with the hardback published in 2015.
Cover illustration: Alexandra Barrio Hilera/EyeEm/gettyimages
Printing and binding: CPI books GmbH, Leck

♾ Printed on acid-free paper
Printed in Germany

www.degruyter.com

Contents

Part II: A Casuistry of Disaster

Isabel Capeloa Gil and Christoph Wulf

Introduction – Hazardous Future: Images and Perceptions of Disaster and the Assessment of Risk

The catastrophic impact of natural disasters in the highly technological socie-ties of the twenty-first century has emphasized the need for us to rethink the assessment of risk within a multidisciplinary framework that does not only look for partial or culture-bound answers but rather approaches environmental is-sues from a systemic perspective. In our current 'risk society', natural hazards are bound up with political decisions and our response to the impacts of catas-trophes is not only culture-dependent but also determined by just how these disasters have been portrayed in the past. Indeed, the security of human com-munities requires a systemic approach that brings together science and tech-nology, economic and social sustainability, political responsibility as well as cultural and historical knowledge.

Modern consciousness has stressed the uncertainties underlying this sub-ject, the threats of war and natural hazard, in a runaway world (Giddens 2002) where every action is framed by terrible risks. In this formidable setting, howev-er, risk was not perceived as thwarting action but rather became a stimulant for creative achievement and an oddly stable doctrine in an otherwise unstable world.

In fact, risk becomes a cultural enabler that draws from a discourse of un-certainty to mediate between the beliefs of the subject and the uncertain odds of the social world. It is then welded into the social fabric by practices of choice, selection and management. Within this process, the untamed irrationality of risk, that which cannot be controlled and is unacceptable to the community, gets transformed into a game of rationally-based probability. This is the feature that reconciles the utter uncertainty of the outside world (be it in nature, in war or in the economy) with the exceptional determination of certain cultures to continue to function within the daunting prospects of environments proving increasingly perilous. Thus, while risks are, in theory, uncontrollable, culture nevertheless seeks to take control of them by creating narratives to insure them-selves through a selection of hazards managed by probabilistic approaches. Culture, therefore, whilst allowing human creativity to thrive by gaming with the odds, will nonetheless support an acceptable level of risks within the safe limits of human social existence. "The perception of risk is a social process. All societies depend on combinations of confidence and fear" (Douglas and Wildavsky 1983, 6). Choice also becomes a risky endeavor, for choosing risk incurs moral, social and political implications. Indeed, this lies at the very root of what Niklas Luhmann named the modern risk society and one where technological development and scientific differentiation feed on the engagement with risk and the acceptance of hazard and actually demanding the existence of risk to thrive and succeed (Luhmann 1988, 38). However, Luhmann also warns that this same society that lives off risky undertakings is threatened by the discourse deeming risk natural. Instead of prompting advancement, this fosters a sort of resilience to the horrible that borders on an apathetic surrender to hazard (Luhmann 1988, 140). Stemming from the research project "Critique of Singularity – The Catastrophic Event and the Rhetoric of Representation" sponsored by the Portuguese Foundation for Science and Technology (2010–2013) and based at the Catholic University of Portugal Centre for Communication and Culture, this book, *Hazardous Future,* represents a collection of articles initially presented at the Hazardous Future Conference, which brought together three interdisciplinary research teams from Portugal (CECC – Research Centre for Communication and Culture, Catholic University of Portugal), Germany (Interdisziplinäres Zentrum für Historische Anthropologie at the Free University of Berlin) and Japan (Korkoro Research Centre, University of Kyoto), in order to discuss the impact of natural catastrophes on the cultural fabric and how the representation of disaster affects risk assessment and the sustainable construction of the future.

It has always been the case that security, desired by all human beings as a basic necessity of life, has only ever existed periodically in human societies. After experiencing the loss of safe living conditions, people have correspondingly always tried to regain those prerequisites essential to living fulfilled lives, both as individuals and in the community. However, whenever we look back at history, it becomes clear this has never been possible for any longer than a brief period of time. Today, not only do natural catastrophes afflict human beings, such as the 2004 tsunami in the Bay of Bengal and Hurricane Katrina, but there is also an increasing number of manmade disasters, such as Chernobyl in 1986 and Fukushima in 2011, which hold long-lasting destructive impacts on our belief in a secure, non-hazardous future. As a consequence of such external influences and also internal social developments, there have been evermore outbreaks of violence that not only destroy everyday life in the regions affected but have also begun to shake and even to silence the "grand narratives" (Lyotard 1984) that envisage a secure and better future. The more open societal development appears to people, the more insecure will their future seem to them within a framework where security is no longer guaranteed either by God or any faith in the "grand narratives". With this loss of a secure and meaningful future, the present day and that happening in the here and now are becoming increasingly important to the lives of human beings. A consequence of this development, in which images and narratives of the future lose their importance as correctives to the present, perhaps also reflects in the way we have come to place excessive value on quantitative empiricism with its inherent assumption that measuring the world is all that is necessary for grasping its reality.

Increasingly now, the future is perceived as something unknown and something that only becomes less terrifying when we begin realizing that even as our knowledge increases our lack of knowledge does not get reduced but rather rises in parallel with our knowledge. The dream of being able to create a secure future for ourselves has never come to fruition. One reason for this stems from how 'not knowing' is normally considered as something negative with demands constantly being made on people to know more than they are actually really able to know. This certainly proved the case in manmade disasters such as Chernobyl or Fukushima. In these cases, there was not only an attempt to ascertain just what caused the disaster but also to identify people to blame, people who supposedly might have prevented the disaster had they somehow been more efficient or better qualified. Trying to identify the reasons for a disaster also nurtures a hope that the means will be found to avoid such a disaster in the future. This approach suggests that the increase in knowledge will make disasters avoidable. This overlooks the fact, however, that disasters are often unintentional side effects of complex social and cultural processes that are generally very hard to anticipate due to their complexity and multi-faceted nature. Thus, in an increasingly globalised world, a world risk society comes into being structures determined by 'not-knowing' the future and the dangers it brings (Beck 2008).

Over recent decades, a professional discourse on risk and hazard has become important as a way of coming to terms with the increasing prominence of natural disasters arising out of the extent of media coverage. This also enables us to engage with the mounting sense of danger regarding the future of mankind. Nevertheless, most of the research done in this field and the policies resulting suffer from a technocratic outlook based on managing technical and organizational resources in order to tackle the structural deficit of the communities affected. They ignore belief systems, historical tradition and representations of disaster within a particular culture. In fact, the portrayal of disaster in literature and the arts is an authoritative source of knowledge for a society and sometimes ends up more important than policy reports in terms of its impact on groups stricken by poverty or tragedy (Giddens 1991).

Aid strategies often perceive communities affected by natural disasters as mere victims, recipients of a one-way stream of aid and not as creative partners in the restructuring of their communities. Furthermore, the technocratic focus tends to lose sight of the role played by the portrayal of disaster in restructuring communal life, in overcoming trauma and in assessing an undoubtedly hazardous future.

Hazardous Future addresses the importance of the image conveyed and mediatized in the aftermath of catastrophe and also the recovery of the affected communities themselves. A variety of perspectives from across the humanities and social sciences, put forward by anthropologists, historians, cultural critics, media experts, psychologists, educators, geographers and other scholars, approach here the different ways in which societies represent disaster. The different chapters focus on geographically diverse situations, looking back to cultural traditions framed by the enormity of past events that provide meaning to risk in the present (such as the San Francisco Earthquake in Isabel Gil's chapter or the Dutch flood of 1953 in Daniela Agostinho's) demonstrating through concrete examples how the representation of earthquake trauma or tsunami fear generates an impact on recent victims. Although this discussion takes place with the benefit of hindsight, it brings together those who create images (journalists, artists) with those who analyze their impact (cultural critics, social scientists and policy makers). The book furthermore asks the following questions: how do modern societies deal with a catastrophic event in terms of the images they make of it and what is the social relevance of this strategy? What are the tropes or metaphors in circulation in the different media representing the event? Are they culturally specific or intercultural and why is this relevant? How does the mediatization of catastrophe contribute to the creation of a literature of disaster?

Indeed, although they are one-off events, the portrayal of catastrophes seems to evoke 'restored behaviors' (Schechner 1981). This is because "Disasters do not simply happen" (Hoffmann and Oliver-Smith 2002, 3). They form part of a system involving the event, a population, a culture. Indeed, disaster happens within a culture that then draws on its traditions and belief systems in order to process the event. This time of generalized crisis, when the discourse of calamity from economics to politics and social upheaval seems to be becoming the new master narrative, proves particularly relevant and timely to engage in a critical appraisal of the rhetoric of catastrophe and discuss its impact and effects upon the cultural and social-political fabric and the media. By means of collaborative research, the book contributes to the sharing of different approaches that produce systemic knowledge, taking us beyond the libidinal indulging in crisis and disaster. *Hazardous Future* is structured around two main topics. The first deals with the intersection of a cultural discourse on modernity, catastrophe and natural disaster. The chapters in the book fully accept that even though the representation of different catastrophes may make use of similar rhetorical tropes, these catastrophes are never identical. However, the articles contest the idea of singularity or uniqueness in the culture of disaster by challenging the narrative behind national images of disaster, i.e. that they remain quite exceptional. Selectivity, repetition, citation, transference, excess, remediatization and self-reflexivity are structural forms constructing a rhetorical framework for the acting out of disaster on the representational level and that also enable it to be worked through. Thus, a discourse of the singularity or exceptionality of catastrophes overlaps with narratives of national identity. This challenge to the idea of catastrophes being unique or singular also deals with the commonalities and diversities in the respective representations of similar disastrous events across cultures.

The articles in *Hazardous Future* begin by asking whether disaster is an anthropological condition – a *conditio humana*. Christoph Wulf contends that catastrophic events are evolutionary drivers of world history. In fact, catastrophe is not only a danger to human life, it was at the root of a major change in human evolution. As evolutionary theory shows, the catastrophic encompasses both creative and destructive drivers. The unique dimension of our contemporary stage of development incorporates how the power of destruction that comes both from war and from environmental damage seems to be leading *Homo sapiens* down a clearly destructive path. The human sciences, Wulf argues, have the ethical potential to allow us to rethink the role that destruction plays in the advancement of human society. First and foremost, however, we must establish what the characteristics of a catastrophe are and how catastrophic events become imprinted on the imaginary of human beings. In recent years, prominent examples of disasters are Chernobyl, 9/11 and Hurricane Katrina and their representation in the media. Among the media images are pictures and films of fear-stricken people fighting for their lives and pictures of people in a state of stunned shock who cannot grasp what is happening to them. There are repeated images of floods and collapsing houses, people in distress and danger who have lost everything and whose future outlook is bleak (Medeiros). Reifications and interpretations of disaster date back as far as Greek Tragedy. Here, people strive to avoid disaster and to this end become enmeshed in the very actions that lead to catastrophes. The limitations of human reason and planning become clearly visible. Examples of this are Oedipus and, 2000 years later, Pozdnyshev in Tolstoy's *Kreutzer Sonata*. In the case of *The Kreutzer Sonata*, disaster emerges out of the blue from everyday events (Resina). Without images and narratives, without being portrayed, a disaster can be neither perceived nor experienced just as any perception requires onlookers and witnesses. Only the horrified eyes of the onlooker enable the disaster stories that then integrate into the collective and the individual memory. The airship "Hindenburg" going up in flames and the Twin Towers collapsing in New York make it quite clear – there is no escape from images of catastrophic events once they have been seen; the onlooker is irrevocably bound to the happening (Gil). Images of catastrophes form in the imaginary to become part of one's body memory and leading to traumas and the forming of groups of people affected in different ways. The media help to create the public opposition that releases new forms of social energy (Mendes). The confrontation with death and the fear of death result in individual and collective traumas. At Fukushima, the therapy designed to help people put into words the images that arise as they process their experiences is described (Kawai). The next article also addresses these experiences in Japan and shows how important the loss of familiar places is for people. Shoko Suzuki takes the Greek concept of *oikos* (household, house or family), modifying its meaning to apply it to the situation in Fukushima, to describe how in *oikos topos* (place) and body become merged and that losing it results in a deep sense of nostalgia (Suzuki).

The second part of the book, "A Casuistry of Disaster" looks into the mediatization of disaster, the narrativization of tragedy and the resilience of affected communities. As both myth-busters and myth-makers, the media produce a sense of insecurity that constrains the perception of both dispassionate onlookers and affected populations when it comes to re-establishing the infrastructure of society. The difficulty of describing what constitutes a disaster is clear when taking the example of the "specter" of Chernobyl. Remarkable in this atomic disaster was the large number of people who, with insufficient protection, set to work in an attempt to stem the emission of nuclear radiation. Why did so many people put their lives and health at risk? What moved them to do that? Did they not know about the danger they were putting themselves in? Many of the myths surrounding this disaster can be interpreted as people being at a complete loss as to what to do and demanding a new understanding of manmade risks and a new way of dealing with them (Alves). The next study of the tsunami disaster in Sumatra in the Indian Ocean demonstrates how cultural attitudes differ when dealing with catastrophes. This is a place where people have always lived with a high degree of risk in an environment threatened by earthquakes, volcanoes and tsunami. Allowing themselves 'to go with the flow' rather than 'fighting the catastrophe' proves a correspondingly typical local response. The singularity of this Buddhist-based attitude is a philosophy of coping that is their very own and representing a clear cultural difference to the behavior of people in Chernobyl (Zaumseil/Prawitasari-Hadiyono).

Aesthetic strategies employed to approach and process disasters have long played an important role. They may be interpreted as attempts to ward off the horror the catastrophe provokes. We find an example of this in Naoko Tanaka's

Performance-Installation "The light thrower". Basing her argument on Virilio's interpretation that recent disasters such as Hiroshima, Chernobyl and Fukushima are products of the technological unconscious (Virilio 2007), Gabriele Brandstetter puts performance into a context in which reality, "sensory awareness" and imagination merge in the movements of the dance in such a way that the horror of these disasters is experienced in the performance images (Brandstetter). We find something similar but with a slightly different slant in Gary Hills' video "Incidence of Catastrophe", an artistic treatment of a disaster inspired by a Blanchot novella. In the selection of scenes from this video, aesthetic and narrative productions and performances of disaster are described and analyzed. This makes clear how language alone proves insufficient when it comes to processing disasters. The traumas that arise in their wake prove very hard to recover from even when repeated efforts are made to deal with them by means of imagination and language (Benthien). Analysis of the element of unpredictability inherent in disasters comes with the examples of the collapse of the Hintze Ribeiro Bridge in Portugal in 2001, 9/11 and the pro-Indonesian military action in East Timor. The focus is on the differences in TV coverage of the three disasters that took place on different continents and on the differing portrayals of death and dying. Differences also become apparent in the ways feelings such as pity, horror, interest, sadness, solidarity and surprise are portrayed (Torres). Looking at the question "What do screens screen?", the next chapter is an analysis of the interplay between image, sound and language in the way fires are presented in catastrophe reports on TV. This analysis includes reflections on phantasms regarding the end of the world (Ilharco). Linked with this, zombies (i.e. the living dead) are often staged to appear, posing a threat to the living while giving them a terrible shock. The return of the dead becomes an important issue here. How are the dead imagined, in what images and performances do they suddenly appear and what is said about them? The author takes the example of *I Am Legend* to analyze the background and significance of the zombification of our society (Holm).

The recurring theme of how the global dissemination of images and discourses of risk and catastrophe combines with culture-dependant portrayals and traditional perceptions in the reassessment of future risk. The next chapter deploys the 2011 Lorca Earthquake to determine whether it is possible to measure disasters and guarantee security and if so how. What can a measurement on the Richter scale tell us? Do we not need different criteria and approaches in order to measure the human suffering caused by disasters? There is no doubt that the culture of sympathy we have developed in the West influences the way the media stage and portray pain alongside the willingness to help and that help actually given (Vieira). A further way of portraying the horror of a disaster is to get individuals to report on their experiences and suffering. Thus, the misery caused by Hurricane Katrina might indeed find more vivid expression in the detailed stories of someone like Creighton Bernette than by any recourse to statistical data and the frequent repetition of short reports (Gonçalves). The final chapter stresses once more how strongly our imaginary, in the form of memories, pre-forms our perception of catastrophes and disaster. Agostinho demonstrates how for many Dutch people the perception of the Big Flood of 1953 does not only bring back memories of the floods of earlier decades. There are also images of places devastated in the Second World War that color the way the current destruction is perceived and intensify the horror (Agostinho). To conclude, the book analyzes hazard and catastrophe in their historical and cultural contexts with the aim of presenting an informed literacy of disaster capable of contributing to sustainable human development.

The organizers of this book wish to thank the hosts of the Hazardous Future Conference, the Fundação Oriente at Arrábida Monastery in Portugal for welcoming the project and graciously providing the venue for the seminar as well as the Critique of Singularity research assistant, Elsa Alves for so devotedly having made sure that no one and nothing was left behind.

References

Douglas, Mary, and Aaron Wildavsky. *Risk and Culture. An Essay on the Selection of Technological and Environmental Dangers*. Berkeley and Los Angeles: University of California Press, 1983.

Giddens, Anthony. *The Consequences of Modernity*. Stanford: Stanford University Press, 1991.

Giddens, Anthony. *Runaway World. How Globalization is Reshaping our Lives*. London: Routledge, 2002.

Hoffman, Susanna M., and Anthony Oliver-Smith. *Catastrophe & Culture: The Anthropology of Disaster*. New Mexico and Oxford: School for Advanced Research Press, 2002.

Luhmann, Niklas. *Soziologie des Risikos*. Frankfurt am Main: Suhrkamp, 1988.

Schechner, Richard. *Between Theatre and Anthropology*. Philadelphia: Pennsylvania University Press, 1981.

Virilio, Paul. *The Original Accident*. Cambridge and Malden, MA: Polity Press, 2007.

Part I: **Disaster (and) Culture**

Christoph Wulf
The Productivity of Catastrophes in Evolution and the Challenge of Manmade Catastrophes

A glance at the history of evolution shows how the Darwinian theory of evolution needs rewriting if one proceeds from today's generally accepted assumption that a celestial body struck the earth 65 million years ago at the end of the Cretaceous period. According to recent calculations, an asteroid with a diameter of 100 meters and a weight of 30 million tons would have released energy equivalent to 3,000 megatons upon impact; this would have led to such high temperatures and extensive firestorms that it would have wiped out not just the dinosaurs but the majority of all life forms of that time. In view of the devastating consequences of such an impact, the question becomes not so much how so many species went extinct as a result but rather just how some forms of life were able to survive this catastrophe. After speculation in previous centuries about a possible asteroid impact, such a catastrophic impact is today seen as largely confirmed not least due to the research of Nobel Prize laureate Luis Alvarez and his son Walter Alvarez.

In the Bottacione Gorge near Gubbio in Italy's Umbria region, Luis Alvarez made the surprising discovery that in the "K-T boundary", a one-centimeter-thick layer that marks the boundary between the Cretaceous and Tertiary periods, the iridium content was 63 times higher than normal. Similarly high iridium content was also found in other locations in Europe and New Zealand. This high iridium content could not be the consequence of meteorite dust distributed uniformly across the earth but instead resulted from the impact of an asteroid. In addition, fossil evidence helped prove that animals died off on a massive scale simultaneous to the iridium being deposited in this layer and leading to the extinction of not only the dinosaurs but also many other forms of life. There was a great deal of evidence that the two facts were interrelated and that both could only be explained as the consequence of an asteroid impact. In the words of Alvarez and his colleagues:

> In brief, our hypothesis was that an asteroid had struck the earth and caused a crater. A large quantity of dust-like material that was thrown out of the crater reached the stratosphere and was distributed around the planet. Until the dust settled over a period of several years, it prevented sunlight from reaching the earth's surface. The loss of sunlight suppressed photosynthesis, causing most food chains to collapse, which resulted in mass extinction (Alvarez et al. 1980, 1105).

According to Hsü (1990, 173), on land, primarily creatures with a body weight of more than 25 kilograms perished. This mass die-off is only understandable if the biology of the various organisms is understood because the doom (or even just the temporary crisis) of one creature represents an opportunity for the other. Furthermore, this always depends on how each individual goes about living its life according to its respective appropriate conditions.

The asteroid impact would have caused so much heat that all forests within 1,000 kilometers were incinerated: firestorms form along with enormous clouds of smoke and dust. The temperature fell and photosynthesis got disrupted. The consequences would have been hypothermia and the starvation of many organisms. The pollution of the atmosphere with nitrogen oxides and "acid rain" led to lung edemas and death never mind how the asteroid's impact may well have triggered earthquakes and volcanic eruptions.

As a consequence of the relatively probable asteroid impact 65 million years ago, scientists then attempted to explain all five major catastrophes with mass extinctions in the Ordovician, Devonian, Permian, Triassic and Cretaceous periods through the quasi-recurrent impact of celestial bodies. This emphasized how the impacts of celestial bodies might have played a central role in the history of life. Astrophysicist Lewis explains such ideas as follows:

> Before life existed there, impacts must have formed simple organic molecules such as formaldehyde and hydrogen cyanide in the ancient atmosphere. These molecules reacted immediately with water, thus forming a great number of molecules that could serve as building blocks for life. The creation of life was thus supported by impacts in earth's early period, just as was the case with other high-energy processes like ultraviolet light, cosmic radiation, and lightning strikes (Lewis 1997, 159).

In contrast to the asteroid impact at the end of the Cretaceous period, these impacts are thus far only rather hypothetical and not subject to conclusive confirmation.

According to Raup (1992, 224), we need mass extinction for our contemporary understanding of evolution and the prevailing principle of probable selective species extinction based on the organism's lack of adaptation. As explained at the beginning, this leads to the need to modify natural selection as originally postulated by Darwin because this fails to explain the diversity of life. On assuming gradual development without such leaps, as postulated by Darwin, the time available for this diversity to arise proves insufficient. As Oeser (2011, 92) points out, there needs to have been species extinction based not on the inability to adapt but on unforeseeable major catastrophes followed by an explosive phase of new development driving the acceleration of evolution. However, such a major evolutionary impetus surely occurred in the wake of the devastating catastrophe at the end of the Cretaceous period. In the opinion of paleontologists, the rapid rise of mammals features among the most dramatic changes taking place in the Cenozoic, the new era of today's life forms. Following the disappearance of the dinosaurs, the mammals initially remained quite small only to then further develop into many forms of rodents and primordial horses, bats and whales and apes and hominids.

Catastrophes have thus played a productive role in evolution. This also reflects in the evolution of hominids. With the emergence of hominids within the scope of differentiation of mammals subsequent to the catastrophe at the end of the Cretaceous period, a life form emerged that for the first time not only has catastrophe to thank for its existence but also generated the ability to actually cause catastrophes. The history of humankind also contains proof that the phrase *natura non facit saltum* which the gradualists following Darwin insisted upon does not hold true. The punctuated character of catastrophes has again and again brought forth developments of life not susceptible to explanation by the gradualist theory of evolution of Darwin and his followers. This insight coming from those who believe in punctuated equilibrium, emphasizing the significance of individual catastrophes, of course does not mean that Darwin's theory of evolution is false but rather demands that his theory of evolution be modified to include the theory of punctuated equilibrium. A number of developments of life cannot be properly explained through natural selection while facets of punctuated equilibrium theories often propose better explanations. For example, geological changes in the earth's interior may have led to major changes on the African continent.

While 20 million years ago Africa was still thickly forested and contained many species and genera of hominids, the situation had already changed substantially by 5 million years ago. Only a few hominid species were then remaining and at some point within this timeframe the human family arose in East Africa. After another 2.5 million years, the temperature dropped to cause massive ice caps to form on the poles. The geological forces changing the face of the African continent came from the dividing line between two tectonic plates extending from the Red Sea in the north to Mozambique in the south (Oeser 2011, 94–95).

Magma streaming upward created the Ethiopian and Kenyan domes, two massive bulges with heights of more than 2,000 meters. These meant that the winds blowing from west to east, bringing moisture to the continent from the Atlantic, were now impeded by a barrier. West of this barrier, hominids could still live in the forests as they had up to that point. East of the barrier, the climate changed considerably. The rains simply did not arrive. Consequently, the forests disappeared with the emerging savannahs no longer providing hominids with their accustomed habitat. In order to be able to live in the savannah, hominids took on the upright gait characteristic of humans. This enabled faster movement over greater distances and facilitated a better "overview", thus also offering improved opportunities for both protection and hunting. As Leakey and Lewin (1992, 110) point out, the motor of evolution is driven by external environmental changes and not by internal competition. Had temperature not fallen 2.5 million years ago, there would have been no evolutionary impetus leading to the adaptation to a dryer environment and hence no Homo species, no people.

The upright gait was also associated with a strong development impetus for the brain and the adoption of stone tools. The much-vaunted properties of the larger brain and technical capabilities did not emerge until 2.5 million years ago. During a long phase in our prehistory, we remained nothing more than apes with an upright gait (Leakey and Lewin 1992, 99).

Within a period of several hundred thousand years, the volume of the brain tripled. In view of the history of humankind, this development proved so very fast that it cannot simply be explained by the gradual development of natural selection. The Australopithecines had brains of 500 cubic centimeters in size. The 2-million-year-old link between the Australopithecines and Homo discovered by Louis B. Leakey in the Olduvai Gorge, *Homo habilis*, had already attained a brain volume of 729 cubic centimeters. In the case of *Homo erectus*, the brain had grown to more than 1,200 cubic centimeters. Out of the conjunction between the upright gait, freeing up the hands, and this explosive brain growth, this "progressive cerebralization" rendered humans superior to all other organisms.

The closest relatives of modern man – the carnivorous svelte *Homo afarensis*, the *Australopithecus africanus*, the herbivorous *Australopithecus robustus*, the East Asian *Homo erectus*, who perhaps already possessed fire, or the Neanderthal, who had a brain even larger than humans – all died out in a geologically short time period of less than 5 million years. According to Oeser (2011, 97–98), only in this way were modern humans able to develop without competition into what they are today: the unchallenged rulers of this world, able to spread throughout it and who, due to their vast superiority over all other organisms, know only a single enemy: they themselves. However, the complex evolutionary process of hominids also requires correspondingly complex models and explanations. Accordingly, the extinction pattern in hominid evolution needs to be similarly complex and extends from Darwinian background mortality to the catastrophic mass die-off of hominid species.

A number of authors assume that the hominids killed animals and other hominids and even contributed to their extinction. For example, in the same sedimentary layers where traces of Australopithecines were located, Raymond

Dart also found baboons with skull injuries, along with Australopithecines with skull injuries, which he deployed to prove the aggressive character of hominids. Nevertheless, the interpretation of these finds remains controversial among experts. The downfall of hominids cannot be interpreted as a consequence of competitive struggle or of long-term background mortality. Richard Leakey assumes a sixth catastrophic extinction event caused not by nature alone but by early human influence. For example, it is frequently presumed that humans were involved in the disappearance of 57 large mammal species in the Americas at the end of the ice age between 10,000 and 12,000 BCE for example.

For the first time in the more than four billion years of earth's history, the only remaining species of hominid evolution seems able to go far beyond the occasional influence on background mortality and cause a mass extinction among the large mammals of the Late Pleistocene period in a very short period of time in the form of a 'blitzkrieg' (Martin 1984; cf. Oeser 2011, 102).

There is also evidence pointing to the extinction of 50 large marsupials in Australia over the period of 100,000 to 12,000 BCE, leaving only four species in addition to the kangaroo. The evidence in New Zealand is even clearer. Moas, large flightless birds similar to ostriches with a height of more than three meters and a weight of 250 kilos, used to inhabit the islands. They were wiped out by Polynesian settlers known as the Maori. More than half a million moa skeletons were found at Maori excavation sites with these birds serving as both food and raw materials.

Such evidence demonstrates how catastrophes constitute an important evolutionary force. They destroy that previously existing to bring forth the new. Without them there would be no development. No life can stand up to their destructive force and this applies both at the level of individual organisms and in terms of the lives of all species. Literally, millions of species have died out. The Earth's crust contains the grave of billions and billions of organisms. Their passing is required for the emergence of new species, new genera, new individuals. Ultimately, life ends in catastrophes. That applies in equal measure both to individuals and collective forms of life. Evolution theory and catastrophe theory are not mutually exclusive; they depend on each other. Like no other living creature, *Homo sapiens* stand out for their killing of their own kind and their potential to annihilate all life on Earth. *Homo sapiens* are thus simultaneously *Homo demens* (Wulf 2013). What is new, however, is that humans today have the actual possibility of destroying both themselves and all other forms of life. What does this insight mean to us? Does it mean that we need to understand catastrophes as central conditions of human life and its development?

Humans today are not just confronted with natural catastrophes in the form of earthquakes, volcanic eruptions, storms, and floods. They themselves hold the ability to generate catastrophes in the form of *war, climate destruction* and *resource obliteration*. What does that mean for our conception of ourselves? Does the anticipation of catastrophes caused by us make it possible to prevent them from happening? Or will we fail to allay the forces of destruction produced by us? How we answer these questions depends on our image of humanity. According to one image of humanity, we have the ability to act reasonably, that is, to anticipate and avoid catastrophe. According to another image of humanity, the driving forces of our action can only be controlled through reason to a limited extent. The history of anthropology features arguments in favor of both of these images of humanity.

The history of violence and the struggle to bring about peace reveals how humans throughout history have had only limited success in creating the lasting, peaceful conditions in the sense of the absence of manifest violence. The recent UNESCO Monitoring Report (2011) again makes this clear. Whilst it has indeed proven possible to nurture and establish peace in large parts of the world, in other regions of the world this has not been possible. In the many regions marked by violent conflicts and wars, a fulfilling human life and the education and development of the coming generation become almost impossible. In spite of the widely accepted view that human life only achieves success without war and violence, seems unable to ensure the corresponding necessary conditions worldwide.

This failure of *Homo sapiens* to achieve peace raises the question of whether or not we are actually able to control the violent potential we produce. Will we be able to avoid again using nuclear and hydrogen bombs or otherwise plunging the world into a catastrophe that would mean its end as we know it? This question takes in the possibilities and the limits of human freedom and human will as well as the ethical foundations of action in hope of attaining this controllability.

Human-caused climate change and the resource destruction associated with the industrialization of the world raises the issue of sustainability. Do we not need to rethink the usage of non-renewable world resources in order to avoid climate and resource catastrophe? Will we be able to do so? To what extent can understanding the situation lead to a change in behavior and how many people will actually attain this necessary understanding? Is it not the case that many of the driving forces of human behavior are of such a quality that they are not susceptible to control through reason and understanding? In human history, there are many examples conveying how wishes for power, desire, and survival often prove such strong drivers of human action that they override any understanding gained through reason. From this perspective, a climate and resource catastrophe would also be unavoidable.

Whilst a great deal indicates that manmade catastrophes are unavoidable, this means we have no ethically responsible alternative to doing everything possible to prevent manmade catastrophes, that is, to actively work for peace and sustainability as well as for education in peace and sustainability (Wulf 1973, 1974; Wulf and Newton 2007).

It would seem no coincidence that the human sciences are concerned with "natural" and manmade "catastrophe" as a central topic with such focuses potentially leading to insights able to fundamentally change our image of humanity.

References

Alvarez, Luis W., Walter Alvarez, Frank Asaro, and Helen V. Michel. "Extraterrestrial Cause for the Cretaceous-Tertiary Extinction." *Science New Series* 208, no. 4448 (6 June 1980): 1095–1108.

Hsü, Kenneth J. *Die letzten Jahre der Dinosaurier. Meteoriteneinschlag, Massensterben und die Folgen für die Evolutionstheorie.* Basel: Birkhäuser, 1990.

Leakey, Richard, and Roger Lewin. *Origins Reconsidered. In Search of What Makes Us Human.* New York: Doubleday, 1992.

Lewis, John S. *Bomben aus dem All. Die kosmische Bedrohung.* Basel: Birkhäuser, 1997.

Oeser, Erhard. *Katastrophen. Triebkraft der Evolution.* Darmstadt: Wissenschaftliche Buchgesellschaft, 2011.

Raup, David M. *Ausgestorben: Zufall oder Vorsehung.* Köln: vgs, 1992.

UNESCO. *Education for All Global Monitoring Report.* The hidden crisis: Armed conflict and education. Paris: UNESCO, 2011.

Wulf, Christoph, ed. *Kritische Friedenserziehung.* Frankfurt am Main: Suhrkamp, 1973.

Wulf, Christoph, ed. *Handbook of Peace Education.* Oslo and Frankfurt am Main: International Peace Research Association, 1974.

Wulf, Christoph. *Anthropology: A Continental Perspective.* Chicago: University of Chicago Press, 2013.

Wulf, Christoph, and Bryan Newton, eds. *Desarrollo Sostenible.* Münster, New York, Munich and Berlin: Waxmann, 2007.

Paulo de Medeiros
Catastrophe Culture[1]

"Damals geschah, was keiner erwartet hatte.
Diesmal musste man eine Katastrophe für möglich halten."
Christa Wolf[2]

"In the war on terror, imminent catastrophe is taken for granted
yet endlessly postponed. The loop between the present and the future is closed."
Slavoj Žižek[3]

"Are we having fun yet?" That seemingly flippant, ironic and cynical question
that helped launch Zippy the pinhead to comic fame back in 1979 was still relatively innocent. In that year, Jimmy Carter was President and Margaret Thatcher
became Prime Minister; the Shah was deposed, and NASA's space program was
at its peak with the delivery of the first space-shuttle, Columbia; there were a
few natural disasters and plane crashes but Egypt and Israel signed a peace-
treaty, the Bee Gees hosted the UNICEF concert with Donna Summer and Abba,
and a nudist beach opened in Brighton. Progress seemed inevitable, Vietnam
was over and the last decrepit colonial empire, the Portuguese, had finally
crumbled. The internet was about to become reality and the hedonistic 90's
were still the future. All in all one might be tempted to think that the time of
catastrophe was in the past, that Europe in particular had managed to survive
the hecatomb of World War II and the perpetual threat of nuclear Armageddon
that was the Cold War. But if there was any euphoria, it should have been tem-
pered by 1986 with the catastrophe at Chernobyl on 26 April. Evidently it was
not really felt, once the scare was over and almost quickly forgotten as perhaps
more time is needed for the consequences to be absorbed. When in 2001 the
Twin Towers were destroyed, officials in the United States government still
claimed that it would have been impossible to imagine such an attack, in spite

---

1 This essay was written while I enjoyed the incomparable intellectual stimulus provided by
colleagues at Wadham College, who welcomed me as the Keeley Fellow for 2011–2012. I would
like to express my gratitude to the Warden and other fellows whose conversations helped me in
the shaping of some ideas.
2 Christa Wolf, „Bücher helfen uns auch nicht weiter". Interview with Evelyn Finger. *Die Zeit*,
23 March 2011. http://www.zeit.de/2011/12/Interview-Christa-Wolf (accessed 8 November 2011).
"Then what no one had expected happened. This time, a catastrophe had to be accepted as a
possibility" (my translation).
3 Žižek, Slavoj, "Catastrophes Real and Imagined". Lacan.com. 28.02.2003. http://www.lacan.
com/zizekcatastrophes.htm (accessed 8 November 2011).

of literary proof to the contrary. That "failure of the imagination"[4] has been completely reversed and if anything we now seem to live in a time in which catastrophe is expected at any moment. Catastrophes have become not only impossible, but they are no longer conceived as having taken place in the past and being some remote possibility in a future no one of us will ever see. Instead, catastrophes have become very much our present, from the devastation in Haiti to the tsunami and nuclear contamination in Japan, and this has given rise to a form of catastrophe culture that needs examining. Both the debasing of the term catastrophe, no longer reserved for end-of-the-world scenarios but even applicable to a burnt dinner, as well as the imagination of the future precisely as catastrophe, are serious and problematic developments that should not be ignored or shrugged off. Perhaps at no other time has catastrophe been so inextricably bound with spectacle and become so political even if always, indeed even etymologically, bound up with dramatic notions. But the linkage between catastrophe and entertainment, even disaster tourism, poses a new constellation of questions for the relationship between culture and society and politics. The predominance of the visual in relation to how the world is perceived and indeed how catastrophes are not only reported but even 'made' demands serious attention to questions of representation that are not simply a debate for learned academic discussion but have a role in plain socio-political terms. In other words, we need to ask ourselves Zippy's question anew, in full awareness that it is just never possible to neatly separate the guilty and the innocent, the deluded and the realistic and that perhaps what matters most is to keep asking. Images of violent confrontations between citizens and the forces of state authority that also pervaded Europe in the 1970s seem lately to have been confined to other regions of the world. However, as the recent photograph by Aris Messinis, depicting the riots in Athens amply demonstrates, such imagery may still strike us as strange and unexpected even if increasingly becoming our daily reality (Messinis 2011).

---

4 The term, in the sense I have in mind here, derives from its use in the "9/11 Commission Report" (22 July 2004): "We believe the 9/11 attacks revealed four kinds of failures: in imagination, policy, capabilities, and management." http://www.9-11commission.gov/report/911 Report_Ch11.pdf (accessed 18 January 2012). As the *Christian Science Monitor* headline the following day read, "Failure of 'imagination' led to 9/11". However, before then it had been used by Adrienne Rich to denounce the first Iraq war: "War comes at the end of the twentieth century as absolute failure of imagination, scientific and political. That a war can be represented as helping a people to 'feel good' about themselves, their country, is a measure of that failure" (Rich 1993, 16).

Responding to 9/11, in an essay entitled "War, Terrorism and Spectacle: On Towers and Caves", Samuel Weber starts by claiming a strikingly new linkage between all three: "*War* and *terrorism* have traditionally been associated with one another, but to link them both to *spectacle* constitutes a relatively new phenomenon and strikes me then as a distinctively contemporary topic" (Weber 2002, 449). Now, even if he immediately does concede that war has always "been associated with pageantry" he still maintains that "never before has what I would call 'theatricalization' played such an integral role in the strategic planning itself". Tempting as it may be to agree with the view that the intrinsic linkage of spectacle with war and terrorism would constitute a novelty, if one looks at the past it should be clear that spectacle in one way or another was indeed always part of both war and terrorism, even if expressions such as the 'theater of war' might be relatively recent – the OED cites Churchill in 1914 as a first occurrence. If one considers catastrophe, and the specific event that animates Weber was both an act of terrorism as well as a catastrophe, the linkage with spectacle ought to be even clearer. Indeed, I would argue, catastrophe always implies a sense of spectacle and even depends on it in order to be properly catastrophic. As regards 9/11 that is perhaps only too obvious: apart from the immense destruction and the immeasurable suffering of those directly involved, for the event to realize its full destructive and threatening potential, it depended to a great extent – from the beginning, that is from its planning through to its execution – on a calculation of its impact as spectacle, whether one has in mind the symbolism of the towers or the incessant repetition of the images of the planes ploughing into the towers served up by global media as it happened and afterwards. Much has been said on the subject and at the moment I would like to turn not to those overcharged images but rather to another image of that catastrophe that works in an altogether inverse manner.

Suppressed at first, Thomas Hoepker's photo of a group of young people seemingly enjoying the beautiful sunshine in the afternoon of 9/11 has evoked much controversy since it was published in 2006 in David Friend's book, *Watching the World Change: The Stories Behind the Images of 9/11*. Interestingly, the publication of the photo in itself was not what started the public discussion but rather a column in the *New York Times* by Frank Rich, under the heading, "Whatever Happened to the America of 9/12?" (10 September 2006). Rich's interpretation of the photo in question, just one of the points in his argument about how the catastrophe had changed America but not in a positive manner, provoked outraged denials by an editor at *Slate* magazine as well as from some of those depicted in the photo. Even Hoepker himself felt the need to explain how he had taken the photograph and why he had withheld publication for so long. The controversy is well known and there is no need to rehash it here.

However, some of the points made are important for an understanding of the concept of catastrophe in the present because they all revolve around time. Had the photograph been taken on any other afternoon, it would be no more than a nice image of a group of young people. Had Hoepker not delayed publication it might just have been lost in the heat of the moment amidst the frightening impact of so many of the other more direct images of disaster. Had Frank Rich not decided to deploy it as a way into questioning what had happened to America *after* 9/11, it might just never have become the counter-icon that it did become.

The way in which Magnum describes the photograph is provocative enough: "Young people relax during their lunch break along the East River while a huge plume of smoke rises from Lower Manhattan after the attack on the World Trade Center". The closeness of the photo to the attack is made clear by the photographer himself as he says that,

> This image happened, in passing, so to speak, when I tried to make my way down to southern Manhattan on the morning of 9/11. [...] The second tower of the World Trade Center had just imploded; estimates of more than 20,000 deaths were quoted and later discredited. Somewhere in Williamsburg I saw, out of the corner of my eye, an almost idyllic scene near a restaurant – flowers, cypress trees, a group of young people sitting in the bright sunshine of this splendid late summer day while the dark, thick plume of smoke was rising in the background. I got out of the car, shot three frames of the seemingly peaceful setting and drove on hastily, hoping/fearing to get closer to the unimaginable horrors at the tip of Manhattan (Hoepker, "I Took That 9/11 Photo", 2006).

It is clear that in the same moment that Hoepker admits responsibility for the image, "I Took That 9/11 Photo", he also takes care to signal that the image itself appeared to him almost by chance, in a side glance, as if not really a product of his trained gaze. It does not matter: even had he staged it on purpose, the image must be read free of its possible intentions just as we must also leave aside what exactly those young people were doing at that precise time.

The apparent serenity of the scene, in spite of the funereal cypress trees, is what stands in stark contrast to the magnitude of the destruction that is merely visible in the column of smoke spreading over the horizon. It is as if the catastrophe were present but at the same time invisible and that certainly must be one the reasons why Hoepker ended up by bagging the photo when, as a senior member of the Magnum cooperative, he had to select which photos from that day taken by a number of colleagues would be used to cover the catastrophe. Hoepker is also ready to point out one of the disturbing aspects of that photograph:

> Four and a half years later, when I was going through my archive to assemble a retrospective exhibition of my work from more than 50 years, the color slide from Brooklyn sudden-

ly seemed to jump at me. Now, distanced from the actual event, the picture seemed strange and surreal. It asked questions but provided no answers. How could disaster descend on such a beautiful day? How could this group of cool-looking young people sit there so relaxed and seemingly untouched by the mother of all catastrophes which unfolded in the background? Was this the callousness of a generation, which had seen too much CNN and too many horror movies? Or was it just the devious lie of a snapshot, which ignored the seconds before and after I had clicked the shutter? (Hoepker, "I Took That 9/11 Photo", 2006).

In that succinct questioning, Hoepker takes us through the traditional questions on photography and realism, photography and documentary truth whilst constantly careful not to compromise himself by a liberal use of question marks and 'seemingly'.

Nevertheless, he also raises some crucial questions: one, that of time, the lapse of several years after the event and the use of the archive; two, the issue of whether media coverage of catastrophe and its representations might have rendered a whole generation numb to them. It is more or less this last point that had been advanced by Rich and had indeed caused the controversy in the first place:

Seen from the perspective of 9/11's fifth anniversary, Mr. Hoepker's photo is prescient as well as important – a snapshot of history soon to come. What he caught was this: Traumatic as the attack on America was, 9/11 would recede quickly for many. This is a country that likes to move on, and fast. The young people in Mr. Hoepker's photo aren't necessarily callous. They're just American. In the five years since the attacks, the ability of Americans to dust themselves off and keep going explains both what's gone right and what's gone wrong on our path to the divided and dispirited state the nation finds itself in today (Rich 2006).

Rich seizes on the political implications of the photograph. It should have been obvious that he is not at all interested so much in what those young people might actually have been doing but rather in the fact that they are, presumably, American. He applies that to levy a heavy charge at the entire nation, while praising it for coping with catastrophe but also accusing it of being too ready to forget and for having come to accept catastrophe as normal: "And so here we are five years later. Fear mongering remains unceasing. So do tax cuts. So does the war against a country that did not attack us on 9/11. We have moved on, but no one can argue that we have moved ahead" (Rich 2006). Even if Rich is still holding on to a liberal notion of progress, he clearly pinpoints the way in which political gains were made from the catastrophe by an unscrupulous administration and perhaps this caused the whole controversy to escalate rather than the photograph itself. To return to Hoepker's first crucial observation, however, it must be said that he again almost denies agency when he says that it was in the process of searching his archive that the photo in question "seemed to almost jump" at him. Nonetheless, he indicates that this was made possible by the passing of time. While the catastrophe was still raging, the photograph was deemed inappropriate to represent the horror; but a few years afterwards, it seemed more important and hence the time lapse must again be seen as one of the decisive elements for interpreting that image. Both the actual time of the catastrophe as well as its future are already contained within it.

One of the key statements on the representation of catastrophe, or disaster, as Blanchot refers to it, would be its relationship to the future or, to be more precise, following Blanchot, the impossibility of imagining the future: "Penser le désastre ... c'est n'avoir plus d'avenir pour le penser" (Blanchot 1980, 7).[5] But this is not true. However much one might think that catastrophe would so affect those experiencing it or thinking about it that they would cease to imagine the future, the opposite proves rather the case. I have had opportunity to reflect on precisely this point in reference to the Lisbon earthquake of 1755 (Medeiros 2005). One of its most eloquent witnesses, the papal nuncio, wrote to Italy at the very moment the catastrophe was still unfurling, having just barely escaped from his collapsing house, to say that the magnitude of the disaster was such that it might take two centuries for Lisbon to recover. That may indeed be a long period of time but implies rather than seeing the catastrophe as the end of the world, the nuncio could well see a future however harsh it might be. What is at stake in Hoepker's photograph is similar, or perhaps even stronger, and that is that even at the precise moment of the catastrophe, the thought of the future is present. Hoepker is right in saying that the photograph raises many questions. He concludes his statements by saying that the photograph is "fuzzy and ambiguous in all its sun-drenched sharpness." Not that the photograph is fuzzy but rather our ability to come to terms with what it implies, namely that we might have entered an era of catastrophe culture, an era in which catastrophe is seen as normal rather than constituting an exception.

Aric Mayer, who covered the devastation of Hurricane Katrina, has also assembled a series of images that are striking for their beauty as much as for the uncomfortable, eerie, feeling they provoke, such as that depicting flooded streets in which the invading waters are completely still, serving as a mirror to reflect whatever manages to still rise above them in a version of an American

---

5 It should be clear that I am not in any way rejecting the importance of Blanchot's thought, merely questioning the applicability of that one lapidary fragment to our present situation. For a detailed analysis of the relevance of Blanchot's thinking for contemporary theory relating to ecological issues, see Timothy Clark's essay on "Blanchot and the End of Nature" (2010).

idyll that is as threatening as it is beautiful. In an essay he wrote about his work, "The Aesthetics of Catastrophe", Mayer reflects seriously on that ambiguous character of his images. Determined to cover the devastation wreaked by nature, Mayer was confronted by the seeming inability of the photographs to depict the horror of what he had seen and thus resorted to an attempt at forcing the viewer to reflect on it. Alert to the way in which the production and framing of images is a key element in politics and in the formation of citizen identity in any given polity, Mayer writes:

> The images of figures struggling in the floodwaters or abandoned without aid were deeply moving, but they spoke only to the immediate plight of those subjects, and to events now passed. Those images captured and conveyed very little of the massive impact of the natural forces that had hit New Orleans, the after effects of which were visible across the cityscape. While the documentary and photo-journalist accounts were troubling to viewers, they were in fact more comforting than the reality of the city, which seemed to defy and elude the available means of media representation (Mayer 2008, 180).

Whereas Hoepker credits chance and the passing of time for realizing how his photograph problematizes the easy ways in which the media representations of catastrophe operate to produce domesticated citizens that accept the inevitability of catastrophe as much as they quickly forget about it, Mayer deliberately set out to provoke such a prick in the general numbness, as he writes towards the conclusion of his essay:

> By neither relieving nor assigning moral culpability, these landscapes are intentionally ambiguous. They are at once beautiful and disturbing. To an audience that is used to having its moral positions clarified and repeated by the media it consumes, this experience of ambiguity in the face of a national catastrophe can be uncomfortable (Mayer 2008, 183).

One way in which the media lulls us into docility and acceptance of a catastrophe culture, of course, is by incessantly showing us far-away disasters that somehow involve us minimally as fellow human beings and which seem more geared to satisfying our hunger for seeing the suffering of others in the relative coziness of our homes whilst forgetting, even if momentarily, the state of perpetual crisis, political even more than financial, that has taken a grip of western societies in the last decade. In the light of such distant cataclysms, we can hardly be moved even to open our purses. The more remote any given catastrophe, the better for such a state of affairs. Furthermore, if the recent catastrophe in Japan has moved us somewhat more than say, the one in Haiti, that has more to do with the fact that we recognize Japan as a developed society very much like our own and are thus better able to feel the threat even while, after the shocking news recede from the television screens, quickly returning to our routines with only a few bothering to inquire about the causes of the current catastrophic state of the world or trying to conceive of ways out of it. It is this seeming acceptance of catastrophe as normal that, if anything, marks a change in our cultural perception of it.

Another way in which we are rendered submissive, even willing, spectators of catastrophe is through the endless reeling out of catastrophe films that not only work towards an acceptance of disaster as normal, but also reassure us that there will always be a future no matter what. Of these, perhaps the best example is the film *2012*, directed by Roland Emmerich, an old hand at such fare, and released in 2009. One of the trailers distributed before the cinema release of the film briefly encapsulates the main points of my argument inasmuch as it focuses on spectacular and massive destruction of all sorts and yet reinforces our wishful thinking that in the end it all will turn out for the best. As the father-hero, Jackson Curtis (John Cusack), dutifully promises at the end of the trailer: "No matter what happens we'll all stay together." Needless to say, critics panned the film even while usually also praising it for its entertainment value. Peter Bradshaw's review in the *Guardian* can be seen as a summation of such sentiment: "This is a wildly over the top anthology of disaster pictures old and new, and Emmerich isn't above recycling other people's ideas. But it's enjoyable and the opening CGI thrill-ride through the collapsing streets of Los Angeles is undeniably good." Needless to say, the film has been a great hit with audiences everywhere judging by its box-office success, with various sources indicating a gross of over 750 million dollars, before DVD sales (over fifty million dollars), on a 200 million dollar budget.

Are we having fun yet? Of course, what makes films such as *2012* so good is that they are so bad. It is not just the pile of cultural and significantly religious references – from the break-up of the covenant between God and man in the fissure of the Sistine chapel and the crumbling of Rio's Christ statue to the ghostly return of JFK as the decommissioned aircraft carrier by that name is the first object to hit the shore in the giant tsunami – that reinforce our numbness. Above all, it is the human drama that the film capitalizes on, with its melodramatic lines reassuring us about the future. Its line about the end being just the beginning is only partly a titillating threat as it also lets audiences believe in their own survival and thus avoid any problematization of the causes for catastrophe.

A recent exhibition at the Freud Museum in London, by the artist Barbara Loftus, entitled *Sigismund's Watch: A Tiny Catastrophe*, conversely, seeks to alert us to the causes of catastrophe and refuses to indulge in a generalized apocalyptic feeling. The exhibition does refer back to the Holocaust, but unlike Blanchot's observations, also modeled on the dominant hecatomb of the twentieth century, this not only imagines a future but represents that future as a form of survival made possible by the reenactment of trauma and its remembrance. Perhaps one could hazard a distinction between the representation of catastrophe in the century just past and our present one: for the twentieth-century, thinking about the future was made rather difficult because of the sheer magnitude of World War II and the constant nuclear threat that the Cold War ceaselessly imposed. In our present, though, after the dissolution of a tenuous bi-polar balance of power, thinking about the future has basically become a thinking of the present as the future; or at worst, a return to the past, evident in the current predilection for the far-right in both European and North-American politics, the rapid dismantling of many of the social gains of the last decades alongside the sense of paralysis that has taken over western governments. The Loftus exhibition is remarkable on a number of accounts especially as it runs counter to that trend. It focuses on catastrophe but not as one generally understands it directly, even if that is its inescapable background of course. Rather, by taking its cue from Siegfried Kracauer, it focuses on a tiny catastrophe, the witnessing by a little girl, the artist's mother, of a struggle between her parents caused by the financial ruin brought on the family by hyperinflation and the rise of Nazism and the smashing of the father's gold watch under the mother's heel. Loftus provides a series of paintings and sketches that revisit that primal scene and reenact it continuously, focusing on the act of gazing and witnessing performed by Hildegard, the little girl: the break up of her world symbolized in the disappearance of the doll she had held as her effigy as well as the couple of bourgeois adults, with her remaining wooden toys scattered and disarrayed, except for the figure of two Chinese works that remains as a constant. The watch, with its multiple meanings, is then held up as a sort of object for reflection that barely escapes fetishization but already points out the need to consider time when reflecting on catastrophe. In a way, Loftus's work both goes back in a process of remembrance and witnessing and moves forward as that involved is rather an act of postmemory in Marianne Hirsch's sense, than any actual remembrance.[6] In the process, Loftus rescues Kracauer's insight from

---

6 Marianne Hirsch's concept of postmemory, presented in her book *Family Frames: Photography, Narrative and Postmemory* (1997), has been widely referenced and debated. In the abstract for her recent article, "The Generation of Postmemory" (2008), Hirsch provides a succinct definition: "Postmemory describes the relationship of the second generation to powerful, often traumatic, experiences that preceded their births but that were nevertheless transmitted to them so deeply as to seem to constitute memories in their own right" (103). The article explores further a number of different approaches to memory studies and her relation to them.

1929 and catapults it right into our own catastrophic present. This is the Kracauer citation that Loftus refers to:

> Man entledige sich doch des Wahns, daß es auch nur in der Hauptsache die großen Geschehnisse seien, die den Menschen bestimmen. Tiefer und dauernder beeinflussen ihn die winzigen Katastrophen, aus denen der Alltag besteht [...] (Kracauer 1971 [1929], 56). [We must rid ourselves of the delusion that it is the major events that have the most decisive influence on us. We are more deeply and continuously moved by the tiny catastrophes that make up daily life].[7]

Perhaps that might prove a way of breaking out of the mesmerization with end of the world scenarios. Feasibly, by paying attention to those tiny daily catastrophes, one might start questioning our acquiescence to catastrophe not only as an event but also as a form of culture in itself. Rather than diluting the sense of magnitude that accompanies a term such as catastrophe and that has been thoroughly degraded in our present in which a less than successful dinner party can be termed catastrophic and without any of the joviality that went with the sixteenth century's use of the word as an euphemism for 'ass', Kracauer's incisive comment calls not for a hedonistic and narcissistic turn into ourselves but rather to an acknowledgment of how damaged our daily lives often are. In the current state of financial and political crisis, perhaps that might also be a way of thinking of ways of breaking the current impasse. Alternatively put, as Boaventura de Sousa Santos recently said, when commenting on yet another round of austerity measures taken by the Portuguese government, "[I]n the light of experience, the recipes of the World Bank and the IMF have not worked out and have only ended in catastrophe and disobedience. However sooner the disobedience, the smaller the catastrophe".[8]

We should keep in mind, however, that the events beginning to reveal fissures in Western polities such as the recent riots in London (August, 2011) cannot be disassociated with other events around the globe and that the majority of catastrophes, whether natural or manmade – a distinction that has almost lost any significance – are still inextricably tied up with Europe's imperial past; a past marked by the conditions of extreme exploitation that still prevail in the guise of neocolonialism and assumed national security concerns throughout

---

7 The English translation provided by Loftus diverges slightly from the published translation by Quintin Hoare, which reads thus: "We must rid ourselves of the delusion that it is major events which most determine a person. He is more deeply and lastingly influenced by the tiny catastrophes of which everyday existence is made up" (62).

8 Comments made in interview with the SIC news program of 14 October 2011. Currently accessible at: http://www.youtube.com/watch?v=G9cnR1x7dVw

Northern Africa, the Middle East and other parts of the globe. 9/11, ten years past, can be seen not as much as the catastrophe that changed everything but rather as the catastrophe that lets us see how enmeshed in catastrophe culture we have all become. To conclude, let me just refer very briefly to the haunting work of Kate Brooks, who after 9/11 has been covering the "theatre of war" that has resulted in its aftermath, whether in Afghanistan, Pakistan, Iraq, Lebanon and other troubled regions. In one of her photographs included in *The Light of Darkness*, Brooks shows a picture documenting the devastating effects of an Israeli raid in Beirut. As opposed to Hoepker's photograph as is possible, in its raw depiction of destruction, it achieves a similar effect. By showing us how deep the craters are, and how completely destroyed the urban space is, the photograph reminds us that the devastation of 9/11, no matter how horrific and symbolic, is neither isolated nor unique. At the same time, by including the figure of a small child in that photograph, walking alone amidst the rubble, Brooks, in a way, also asks us to consider the more invisible, tiny, yet highly significant catastrophes Kracauer warns about. As we look at that photograph – or others, such as the one she chose as her book cover, depicting two children standing alone inside ruins, barely illuminated except for the sunlight coming through the broken windows, indeed, in the light of darkness – we are, or we should be, forced to relinquish our comfortable positions, the blasé carelessness embodied in the question of "are we having fun yet?" and start questioning how it was possible that we came to see catastrophe as a normal part of our culture.

References

Blanchot, Maurice. *L'Écriture du désastre*. Paris: Gallimard, 1980.
Clark, Timothy. "Blanchot and the End of Nature." *Parallax* 16, no.2 (2010): 20–30.
Friend, David. *Watching the World Change: The Stories Behind the Images of 9/11*. New York: Farrar, Straus & Giroux, 2006.
Hirsch, Marianne. *Family Frames: Photography, Narrative and Postmemory*. Cambridge, MA: Harvard University Press, 1997.
Hirsch, Marianne. "The Generation of Postmemory." *Poetics Today* 29, no.1 (2008): 103–128.
Kracauer, Siegfried. *Die Angestellten*. Frankfurt am Main: Suhrkamp, 1971 (1929).
Kracauer, Siegfried. *The Salaried Masses: Duty and Distraction in Weimar Germany*. Trans. Quintin Hoare. London: Verso Books, 1998.
Mayer, Aric. "Aesthetics of Catastrophe." *Public Culture* 20, no. 2 (2008): 177–191.
de Medeiros, Paulo. "De escombros e escumalhas." In *O Grande Terramoto de Lisboa: Ficar Diferente*, eds. Helena Carvalhão Buescu and Gonçalo Cordeiro, 243–264. Lisboa: Gradiva, 2005.

Rich, Adrienne. What Is Found There: Notebooks on Poetry and Politics. New York: W. W. Norton, 1993.

Weber, Samuel. "War, Terrorism, and Spectacle: On Towers and Caves." *The South Atlantic Quarterly* 101, no.3 (2002): 449–458.

Internet:

Bradshaw, Peter (2009). "Roland Emmerich Returns. Globe Shudders." *Guardian*, 13 November 2009. http://www.guardian.co.uk/film/2009/nov/13/2012-film-review (accessed 18 January 2012).

Chaddock, Gail Russel (2004). "Failure of 'imagination' led to 9/11." *Christian Science Monitor*, 23 July 2004. http://www.csmonitor.com/2004/0723/p01s03-uspo.html (accessed 18 January 2012).

Emmerich, Roland, dir. (2009). *2012*, Columbia Pictures. Sony, DVD. 29 March 2010. Trailer: http://www.sonypictures.com/movies/2012/ (accessed 19 January 2012).

Hoepker, Thomas (2001). "USA. Brooklyn, New York. September 11, 2001. Young people relax during their lunch break along the East River while a huge plume of smoke rises from Lower Manhattan after the attack on the World Trade Center." http://www.magnum photos.com/C.aspx?VP3=ViewBox_VPage&VBID=2K1HZOO6O4PBQ&IT=ZoomImage01_V Form&IID=2K7O3RK0762&PN=2&CT=Search (accessed 8 November 2011).

Hoepker, Thomas (2006). "I Took That 9/11 Photo." *Slate*, 14 September 2006. http://www. slate.com/articles/arts/culturebox/2006/09/i_took_that_911_photo.html (accessed 8 November 2011).

Kean, Thomas H. et al (2004). "The 9-11 Commission Report. Final Report of the National Commission on Terrorist Attacks Upon the United States." *Official Government Edition*, 22 Juli 2004. http://www.gpoaccess.gov/911/ (accessed 18 January 2012).

Loftus, Barbara (2011). *Sigismundarbara*. London: Vers, London: Phillip Wilson. "Exhibition at the Freud Museum, London, Curated by Monica Bohm-Duchen, 5 October to 13 November 2011." http://www.freud.org.uk/exhibitions/74347/sigismund47/sigismund%C4%81's-watch-a-t (accessed 8 November 2011).

Messinis, Aris (2011). "Masked Attacker." Guardian Eyewitness Photos. http://www.guardian. co.uk/world/picture/2011/jun/29/greece-euro-eyewitness (accessed 08 November 2011).

Rich, Frank (2006). "Whatever happened to the America of 9/12?" *The New York Times* 10 September 2006. http://www.nytimes.com/2006/09/10/opinion/10rich.html?pagewanted= all (accessed 19 January 2012).

Santos, Boaventura de Sousa (2011). "Interview with SIC news program." 14 October 2011. http://www.youtube.com/watch?v=G9cnR1x7dVw (accessed 19 January 2012).

Wolf, Christa (2011). "Bücher helfen uns auch nicht weiter." Interview with Evelyn Finger, *Die Zeit*. 23 March 2011. http://www.zeit.de/2011/12/Interview-Christa-Wolf (accessed 8 November 2011).

Žižek, Slavoj (2003). "Catastrophes Real and Imagined." Lacan.com. 28 February 2003. http://www.lacan.com/zizekcatastrophes.htm (accessed 8 November 2011).

Joan Ramon Resina
Tragedy, the Neglected Origin of Catastrophe Theory

In *The Birth of Tragedy*, Nietzsche wrote: "If we could imagine dissonance become man – and what else is man? – this dissonance, to be able to live, would need a splendid illusion that would cover dissonance with a veil of beauty" (143). Nietzsche's allusion to Schopenhauer's philosophy in this reference to the blinding effect of the will is transparent, as is his esthetic modulation of the illusion, cast in the service of harmony and form; or in the young Nietzsche's terms, of Apollonian rationality as the strongest enticement to individuation. Nietzsche implicitly raises the question whether human life is possible when deprived of beauty or the hope of beauty. His variation on Schopenhauer's idea of the will being the source of metaphysics takes illusion farther in that it posits an aesthetic rather than a moral principle as the force unifying the subject. On this view, the subject's apparent unity would be as illusory as the superficial harmony disguising its inner dissonance. Failure of the illusion or a breakdown of the unity into its dissonant components entails catastrophic consequences for the subject. Indeed, catastrophe has no better definition than the lifting of the mist that blinds us to our inconsistency while protecting us against the collapse of our sense of embodying an autonomous self capable of opposing valuations and moral achievements to the disaggregating force of natural destiny.

Nietzsche's text is the place where the original meaning of catastrophe as dramatic resolution meets the modern sense of the term as an event fraught with large-scale consequences originating beyond human volition or control. It was in the eighteenth century that the term "catastrophe" began to be used in the sense of natural disaster. Until then it retained the theatrical meaning of an ending or resolution, as can be seen in Rabelais, although he applied it by extension outside the theater. But even in the eighteenth century the term preserved, side by side with the new connotation, the sense of unhappy ending reminiscent of the dénouement in classic tragedy. Michael O'Dea asserts that this acceptation disappears at the end of the eighteenth century and is excluded from current usage (2008, 38). It seems an exceedingly confident assertion. Unfortunately, semantic history is not as cut and dry as that. Certainly, most contemporary literature on catastrophe shuns the original use of the term as a sudden and often unexpected resolution of a personal entanglement. But the painful resolution of an untenable or unjust situation based on unseeing arrogance, a resolution that constituted the essence of tragic knowledge, remains within the compass of the modern sense of the catastrophic. We find it, for instance, in the warning that ecological catastrophe may follow from unwillingness to change our depredatory lifestyle. Thus I propose that the ancient meaning of "catastrophe" did not become obsolete after the eighteenth century but retains some of its semantic virtue, designating the conjunction of human will with the manifestation of a force that overpowers it, as if the individual's existence suddenly appeared in the light of an alien interpretive principle.

Among historians there is disagreement about the origin of what we might call a theoretical – rather than theatrical – approach to catastrophe. While Anne-Marie Mercier-Faivre and Chantal Thomas place the invention of catastrophe in the eighteenth century, claiming that before that time religion rather than secular theory dealt with natural upheavals (2008, 9), Alessa Johns acknowledges that "a scientific approach to catastrophe can in fact be traced to ancient times and was constantly vying with a religious one" (1999, xviii). While her book studies exclusively eighteenth century case histories, she admits nonetheless that this period "only shifts the balance between physical and metaphysical explanations of catastrophe" (Johns 1999, xvi), but even ancient theories "were rational, systematic interpretations based on observation to counter what were perceived as superstitious explanations of calamities" (Johns 1999, xvii). But regardless of how the concept is historicized, catastrophe is usually understood as a predetermined event whose approach is unnnoticed until it breaks upon unsuspecting victims. Michael Howard, in his book *Liberation or Catastrophe?* (2007), revisits Theodor Adorno's and Max Horkheimer's basic theme in *The Dialectic of Enlightenment*, when he proposes that Fascism was an unintended consequence of the unfolding of rationality and scientific progress. This process is still going on and the destruction of traditional communities in the Third World under the impact of industrialization is "creating masses of uprooted peoples desperate to escape to wealthier lands, and ripe for militancy if they cannot escape" (11). Given the failure of the two political children of the Enlightenment, communism and liberalism, to deliver on their promises, he predicts another catastrophic turn of events, this time in the Third World. "What is there now left, if liberal capitalism continues to result in misery for the masses, except some form of Fascism?" (12). Howard's is a sober projection on the dotted lines of the future. Others are less restrained. Futurology easily becomes fertile ground for catastrophism. Guillaume Corvus not only presents a whole palette of possible catastrophes but even theorizes their convergence and interaction to produce within a very proximate temporal horizon a radical materialization of Oswald Spengler's prediction of the passing of the West. He talks about an explosive cocktail of nightmares, each of which triggers or exacerbates the others. It is no longer the case that catastrophe approaches unobserved, since the changes leading to it are well underway and perfectly detectable, but the elements of blindness and impotence are nonetheless present, casting a strong sense of fatalism on the outlook for the next decades. Authorities are impotent to change the course of events. Even when they try, they fail to steer the world away from catastrophe (2004, 21). Most people advance relentlessly on the path to disaster. "We are moving toward a de-realization, the loss of the sense of the reality of what happens" (21). Far from being dead, the myth of progress is doing very well, thank you, especially in the emerging countries of the South. And it is this relentless forging ahead, accompanied by what Corvus calls "discrete metamorphoses of structure" that is leading inexorably to a "threshold of rupture," that will bring the downfall of civilization as we know it (19). Less pessimistic, Vaclav Smil remains unconvinced, "despite the enormous challenges we face, that our civilization will be soon transmuted into a defunct heap" (2008, 1). While admitting that the successes of Western culture mislead us into believing that we are in charge of history and projecting a great increase of painful events in the next fifty years, he reminds us nonetheless that the future is not preordained (251). Our adaptive capacities, he says, allow us to cope with change and are capable of turning the greatest crises into resolved challenges.

Whether human made catastrophes are on the rise as a result of expanding modernity or express the exhaustion of modernity, they are linked to vanishing illusions of human autonomy and planned agency. If awareness of impending catastrophe should lead, according to Jean-Pierre Dupuy, to invert the Cartesian dogma, taking the systematic doubt as an occasion not to suspend but to trigger decisions (Walter 2008, 319), then something like a post-Nietzschean willingness to live dangerously may be inferred from his observation. Or, to remain with Dupuy's more literal sense, the very withdrawal from a dystopian future should plunge us into action. Because catastrophes are often disbelieved until they hit and only then appear as inevitable (Dupuy 2008, 488), he proposes an enlightened catastrophism, which would react upon a future that becomes avoidable to the extent that it is discerned and experienced under the form of an unwelcome prediction.

Of course, this position remains embedded in the basic assumption of modernity, namely the abiding resolution of meaningful action into the humanly knowable; more precisely, the delimitation by reason of the area that is accessible to purposive action. Beyond that area lie magic and superstition as a receding realm of futility. In face of catastrophic prospects, futurology encourages avoidance, switching tracks in order to produce through an inventive rearrangement of the features of the present an alternative to the future that persistence would bring to pass. François Walter asks pointedly: "But will it be truly possible – for the first time in history – to build a social future on the basis of an anti-project?" (2008, 320). Tepid response to climate change, for instance, suggests how difficult it is to motivate through negative reinforcement. But another problem with the way societies respond to menaces that lurk in the future lies in their perception. According to Walter, terminology reveals the perception and classification of disruptive events. "Disaster" rather than "catastrophe" is the term chosen by government and relief agencies to refer to disruptive events with non-human causes. "Disaster" suggests a kind of event that happens independently of the human will, while "catastrophe" seems to be not only suffered by humans but also caused by them. Perhaps for this reason, disasters are thought to render human action moot and to allow only for divinatory or, more exactly, hermeneutic dispositions. "Disasters," argue Jane Anna and Lewis R. Gordon, are irrational phenomena from the deep, knowable not through direct manifestation but as "symptoms of more subterranean forces" (2009, 117). These forces are only recognizable through their signs, and it is mainly the inability to read those signs that leads to catastrophe. In other words, human induced catastrophe is the consequence of failure to heed the signs of impending disaster. Thus catastrophe adds a semiological layer to disaster by becoming one more sign in its continuum (8).

Thus, under certain conditions, disaster can trigger catastrophes. Climate change may be the ultimate reason for historical turning points, geopolitical shifts and large-scale civilization failures, as David Keys shows in his book *Catastrophe*. Keys seeks to explain the unraveling of the past in relation to causes outside of human agency; for him it is not a matter of reading the signs of the time and practicing some form of avoidance, since the historian can only point out the dangers but not predict if any of these will materialize and with what consequences for the course of human history (1999, 305). Certain is only that catastrophes have occurred in the past and that there is evidence, scientific as well as statistical, suggesting that they will recur in the future. Walter concurs: "the only certainty is that of a constant threat manifesting a new relation to the world."

Seeing the world as a dangerous place is surely nothing new, and yet catastrophe theorists often depict the present as especially conscious of catastrophe. Corresponding to this attitude, says Walter, the "precaution principle" has become "the vulgate of deciders" (2008, 334). The sentiment of impending catastrophe and the associated unwillingness to risk may be felt more acutely in the wake of what we might call the "disenchantment of modernity," a term that appears redundant only if we forget that modernity fetishized progress and invented a scientistic mythology. Based on local but ascertainable improvement, modernity hinges on the belief that the future will always be brighter than the past and science will remove the obstacles on the way to happiness, including – at the height of scientific arrogance – the formidable obstacle of death. As Dough Cocks explains, this is a dangerous myth, because it leads people to believe that all problems can be solved through the application of Cartesian-Newtonian science and the predictability arising from the discovery of universal laws governing all natural phenomena (2003, 212).

But certainty of recovering paradise on Earth has been exceptional in history, and its dissipation in our time may indicate a return to metaphysical normality, i.e. to a form of expectation that, inscribed within the limits of certain pain, conceives of social life as a fissured landscape prone to shaking. If this were true, catastrophism would be the norm rather than the exception. The reason it reappears as a burning issue now may be due to the collapse of the scientist utopias and the rise of a new skepticism. By comparison with current threats of ecological devastation, extensive famine, massive migration, political destabilization, financial tsunamis, and the reemergence of religious fanaticism, the second half of the twentieth century looks in retrospect like a long period of relative tranquility and unsurpassed prosperity that everything suggests is now over. Contrast between a time in which it was possible to look forward confidently to expanding rationalization and the current defaulting of yesteryear's illusions of progress sufficiently explains the spread of catastrophic apprehension.

Most forecasts insist on the contingency of catastrophic events, while demanding intervention either in the form of action or restraint from action. Thus, catastrophe theory combines chance and responsibility, in essence underwriting the precaution principle. But if chance, the gambling with possibility, opens the dimension of uncertainty and risk, insistence on responsibility safeguards the social requirement of guilt in the absence of its metaphysical necessity. Jean Baudrillard detected the incongruence, longing for a reinforcement of the reality principle and the retreat of the scapegoating principle that technocratic society inherited from its religious predecessors:

> What a liberation it would be for the human mind to recognize [...] catastrophes themselves as natural, i.e. spontaneous, without the intervention of artifice or anyone's will (and certainly not the will of God!). But such is the human mind: being itself artificial, it always needs to impute things to minds or causes. Catastrophes never seem marvelously natural to it, never appear in their fateful simplicity. It wants to be the cause of all these misfortunes and throws itself into this heroic superstition (1996, 48–49).

The need for heroics inheres in the notion of catastrophe. Were catastrophes left entirely in the domain of chance, futurology would be thoroughly discredited. More importantly, by turning its back on the classic notion of catastrophe, theory would be neglecting the connection between efforts to circumvent catastrophe and their eventual downturn when planning spirals into calamity. The classic notion of catastrophe concerns an epistemological event opening upon a panorama of pain. The closest one comes to that classic meaning in the present is in the numerous interpretations of the Holocaust as the twentieth century catastrophe par excellence. This event fits the definition of catastrophe in that it was a manmade calamity of vast proportions, whose victims were loath to admit the evidence of what was happening until it was too late. Regardless of the information that states and some individuals may have had, the fact that the world was unsuspecting of the full scope of the Holocaust until the camps were liberated was tragic in that the massive failure of humanity became imputable as moral guilt *a posteriori*. It was only after the war that the gradual, piecemeal breakdown of human sympathy that had begun long before the conflict could be understood in all its tragic proportions. Yet tragedy is what catastrophe theory rarely brings into its compass. The most likely reason for the oversight is that techno-scientific civilization has now advanced to the point where statistics and the calculation of probability have displaced the messy realities from the logic of human affairs. To the extent that catastrophe theorists appeal primarily to those bodies (governments, industry, consumers, the scientific community) that constitute the foundation of modern society, they credit them with inherent rationality. And yet, prominent among the crises of the present is the crisis of the political. Could it be that the crisis of politics comes from its losing the sense of tragedy? From Athenian democracy through the *signorie* of the Renaissance to the absolutist court of Louis XIV, politics was always pervaded by a sense of the tragic. What were the rituals of public execution if not a manifestation of the pain on which political legitimacy was founded? Is it coincidental that politics began to founder in the century that saw the emergence of an ampler and more general meaning of catastrophe, the century when government took upon itself the obligation to guarantee the pursuit of happiness as an inalienable right? And is not the semantic distance minimal between the right to pursue an object and the right to the object itself? What happens when government is expected to deliver happiness and political parties, vying with each other to sustain the illusion, cannot afford to admit the inevitability of pain? What is the consequence of publicly discarding tragic knowledge?

In his essay on Nietzsche, Peter Sloterdijk suggests that present-day delegitimation of institutional politics is caused not by the greater or lesser mediocrity of politicians but by the appearance of a rift in the concept of the political. The yawning gap in political rationality requires the supplement of a darker concept accounting for "the hidden ecology of universal pain" (1989, 76). Transferring Nietzsche's Apollonian and Dionysian principles to the political context, Sloterdijk proposes that the very attempt to construct socially endurable situations (i.e. variations on the utopian horizons of a science-governed society) brings about the proliferation of suffering through precisely the construction of what can be endured (1989, 77). If he is right to suggest that the darkening of the political landscape over recent decades is due to the emergence of a need for what he calls an "algodicy," an account of universal pain, then catastrophism is not so much the impression of increased exposure to disaster as the revealing of a hidden side to the optimism of modernity.

Modernity based its legitimacy on the promise of a steady diminishing of pain and on occultation of the growing servitude to technical organization of life. In a post-Nietzschean, i.e. Heideggerian critique of modernity, catastrophe is inscribed in *Dasein's* relation to Being, a relation that includes inquiry as one of its possibilities (Heidegger 1962, 27). Because Being is what *Dasein* inquires into, it is at once that which is closest to *Dasein* and that which is most remote in never presenting itself as an object to consciousness. Being, as the Being of entities, is the Dionysian force responsible for the fragmentation of the world (the god that is torn apart at the height of ecstasy). Through its sacrifice and self-effacement, Being allows the emergence of knowledge structures as so many Apollonian strategies for the containment of pain. Such structures become institutional discourses, the *doxa* that constitute reality as experienced when life is shielded from the deeper truth through institutional definitions "in accordance with the criteria of endurability and predictability" (Sloterdijk 1989, 76). Catastrophe is the upsetting of such criteria, resulting in the confrontation of the universality and inevitability of pain and the attendant ruin of received structures of knowledge.

Twenty-five years after the publication of Nietzsche's *The Birth of Tragedy*, Tolstoy produced one of the most poignant analyses of catastrophe in *The Kreutzer Sonata* (1899). In a modern replica to Plato's *Symposium*, where revelers at a drinking party compete in defining love, Tolstoy stages a conversation on the subject in the casual reunion of travelers on a train to Moscow. Love is discussed in the usual romantic clichés until a passenger rebuffs the blithe talkers, then launches into a confession of how his life plunged into catastrophe through the revelation of the dark side of human desire. "We did not comprehend that this love and hatred were one and the same animal passion, only with opposite poles. It would have been horrible to live in this way if we had realized our situation; but we did not realize it and did not see it" (Tolstoy 1957, 62). Unconsciousness is what allows the continuity of a brutal life under the aspect of the ordinary. Charged between the two poles of love and hatred, this unendurable existence is made endurable by the socially stabilizing form of matrimony, hiding all the while the abyssal reality through intoxication with romantic banalities and erotic dreams.

Under these conditions catastrophe is likely, though by no means guaranteed. It retains the contingent character underscored by most theorists: "Thus we lived in a continual mist, not recognizing the situation in which we found ourselves. And if the catastrophe which overtook us had not occurred, I should have continued to live on till old age in the same way, and on my deathbed I should even have thought that I had lived a good life" (63). The poignancy of Tolstoy's account is that the ordinary is catastrophic, that catastrophe has always already occurred, and so, what we call catastrophe is merely the dramatization of a condition, the tipping point at which the tree hollowed by rot crashes down. Catastrophe is thus not the thing revealed but the revelation itself. It cannot be objectified and cast in the mold of any science because it fails to conform to science's requirement of predictability and endurability, i.e. regularity and universality. Not even in its most recent incarnations can science deal adequately with catastrophe because, while there are disasters to prevent or mitigate, catastrophe cannot be rendered specific to any of the discourses of applied science. It is not an object but an event, and a self-cognizant one at that. Says Pozdnyshev:

> These new theories of hypnotism, mental disease, and hysteria are all an absurdity – not a simple absurdity, a vile and pernicious one. Charcot would have infallibly said that my wife was a victim of hysteria and he would have said that I was abnormal, and he probably would have tried to cure us. But there was no disease to cure (63).

No disease because the malady was rather metaphysical, a Schopenhauerian addiction to illusion fed by the will. Pozdnyshev suggests that what befell him, killing his wife, could have happened to any other man. The fake self-assurance that edges him on to take the risks that eventually undo him is the normal condition in a society built upon an abyss of passion that is disavowed. Life, says Tolstoy, is always building up to catastrophe but the latter occurs only at rare moments of self-recognition in the sudden estrangement from self. The moment when *Dasein* comes into the space of discovery, of *Unverborgenheit*, is always destructive. If forgetfulness of Being is inseparable from *Dasein's* mode of existence, remembering Being is destructive of *Dasein's* epistemological categories, since the latter are no longer issuing from the confrontation with that which they seek to apprehend. Being forever defeats knowledge.

Pozdnyshev lives with the accepted notions of love and matrimonial decorum, all the while building up the tension that will explode and release itself as soon as love tips into violence. Uxoricide is only an incidental aspect of the crucial episode. The real event is the lifting of the veil that keeps the husband ignorant, triggering the full, unmediated self-discovery. "I cannot say that I knew in advance what I was going to do, but at the instant I did anything, and perhaps a little before, I knew what I was up to – as if for the sake of being able to repent, of being able to say to myself, 'I might have stopped.' … But this consciousness flashed through my mind like lightning and was instantly followed by the deed. The deed made itself real with unexampled clearness" (Tolstoy 1957, 112). The blaze of consciousness is there precisely to illuminate the deed, much as in Benjamin's well-known metaphor "light flashes up from strange sources as if fuelled from magnesium powder" and imprints a memory as if on a photographic plate (1970, 115). Consciousness is there, says Tolstoy, to ground the emotion of guilt, which is founded on knowledge that the act was both necessary and avoidable, determined and yet preventable through its reflection in the mind and its recognition therein.

But what happens when action and knowledge are not simultaneous, when knowledge throws light on the act long after it has occurred? For Baudrillard, "prophesying catastrophe is incredibly banal. The more original move is to assume that it has already occurred" (1996, 68). This is true literally, for it is exactly the move made by Greek tragedy, where the term "catastrophe" originates. There it was literally the end of the play. An end, as we know, that triggered catharsis, releasing the emotions of terror and pity. And this coincided with anagnorisis, the moment of recognition of misfortune undergone. But placing catastrophe at the beginning, far from relieving us of responsibility, as Baudrillard believes (69), merely makes it retroactive. Oedipus comes upon the meaning of his actions long after the fact through unexpected revelation. His life is the one riddle he cannot solve by his own lights. By the time knowledge dawns on him, the action is no longer revocable. He can only suffer and grow into the darkness of his tragic knowledge. He loses the lucent world of appearance and enters a shadowy one. In Karl Jaspers's words: "Wherever the tragic sense appears, something extraordinary is lost: the feeling of security without the shadow of tragedy, a natural and sublime humanity, a sense of being at home in this world, and a wealth of concrete insights" (1952, 33). Oedipus and Pozdnyshev have this much in common: the dissipation of the illusion of security and harmonic exchange with their fellow men, and therewith the breakdown of arrogance based on social position. But in placing action in the pre-history of drama and focusing on revelation as the stuff of tragedy, Sophocles and Tolstoy show that it is knowledge, not action per se that clinches destiny. Again Jaspers: "Genuine awareness of the tragic […] is more than mere contemplation of suffering and death, flux and extinction. If these things are to become tragic, man must act. It is only then, through his own actions, that man enters into the tragic involvement that inevitably must destroy him" (42). Man acts and thereby gets entangled in a metaphysical struggle between the Dionysian life force that knows no restraint and the Apollonian striving for perfection and repose. The tragic struggle must issue in catastrophe that reverses the correlation of forces at the moment of highest achievement of form. Jaspers explains it as follows:

> What will be ruined here is not merely man's life as concrete existence, but every concrete embodiment of whatever perfection he sought. Man's mind fails and breaks down in the very wealth of its potentialities. Every one of these potentialities, as it becomes fulfilled, provokes and reaps disaster (1952, 42).

Seen from a moral point of view, the process is one of accumulating responsibility. The more a person or society advances on the path of control of its internal and external environment, the more it collects guilt from the binding of energy into a few selected purposes through repression of myriad potentialities. At some point the built-up guilt punctures the restraining tissue and social anomie sets in. In Baudrillard's words, "Every catastrophe bursts the abscess of collective responsibility. Our systems secrete such a charge of floating responsibility that it condenses from time to time like static electricity in lightning, with accidents or catastrophes providing the spark" (Baudrillard 1996, 48). Scott Fitzgerald put it more poetically in his account of the disillusionment that comes with maturity in the wake of unbound aspiration and dreams of a golden youth:

> I began to realize that for two years my life had been drawing on resources that I did not possess, that I had been mortgaging myself physically and spiritually up to the hilt. What was the small gift of life given back in comparison to that? – when there had once been a pride of direction and a confidence in enduring independence (1956, 72).

Catastrophe may well be defined as the moment of realization that one, we, society, have been living on borrowed time and that the pride and confidence in the endurance of that state of affairs is washed away in an instant that steals upon us unexpectedly. The moment the crack opens on the Apollonian surface of our lives, there is no going back to the protective womb of illusion. We are out of the cocoon and must cope with the knowledge that "the natural state of the sentient adult is a qualified unhappiness" and that our noble strivings "to be finer in grain than you are" only add to the unhappiness (84).

There is more despair than resignation in Fitzgerald's words. Western society has also reached the point of a crack up, which makes it vulnerable to its own delusion and explains the foundering of political techno-utopias, including the dream of a democratic global society that would, in Frances Fukuyama's imagining, have brought the end of history. He saw the end of the communist utopia as the triumph of the liberal utopia, rather than the end of political utopias altogether. Even as Hegelian completion of the embodiment of the Idea in worldhistoricality, that is, even as the abolition of transcendence, Fukuyama's proclamation of triumphant immanence was flawed. Nietzsche's rather than Hegel's vision was pertinent. Conceiving life as radical immanence, Nietzsche saw it as a perennial dance between the extremes of form and dissolution. Thus, the triumph of Western liberalism was the onset of its decline. Because Nietzsche did not accept any metaphysics of redemption, there was to be no resolution of the dialectic, no ultimate synthesis, no end of history. History itself was no more than a protective skin to keep us from feeling the terror of immediacy. It was, so to speak, one aspect of culture as the perpetual play between the Dionysian fusion of pleasure and pain and the Apollonian striving after form that for a fleeting moment makes the unendurable endurable. The individual or culture that has looked into the tragic and accepted responsibility for the whole of life does not transcend guilt but rather "accepts danger and that inescapable nexus of guilt and doom implicit in all true action and accomplishment in the real world" (Jaspers 1952, 96). And this not just because all achievement is doomed, but because at the very moment of catastrophe, from the wrecked dream of aloofness from universal suffering, rises a Dionysian liberation from the illusions of idealism and the search for culprits. Through the pangs of unendurable pain, the self-blinded Oedipus turns into Oedipus the seer.

References

Baudrillard, Jean. *Cool Memories II*. Trans. Chris Turner. Durham: Duke University Press, 1996.

Benjamin, Walter. *Berliner Chronik*. Frankfurt am Main: Suhrkamp, 1970.

Cocks, Doug. *Deep Futures: Our Prospects for Survival*. Montreal and Kingston: McGill-Queen's University Press, 2003.

Corvus, Guillaume. *La convergence des catastrophes*. Paris: Diffusion International Edition, 2004.

Dupuy, Jean-Pierre. "Notre dernier siècle." In *L'Invention de la catastrophe au XVIIIᵉ siècle: Du châtiment divin au désastre naturel*, eds. Anne-Marie Mercier-Faivre and Chantal Thomas, 480–494. Genève: Droz, 2008.

Fitzgerald, Francis S. "The Crack Up." In *The Crack Up*, ed. Edmund Wilson, 69–84. New York: New Directions, 1956.

Fukuyama, Francis. *The End of History and the Last Man*. New York: Avon, 1993.

Gordon, Jane A., and Lewis R. Gordon. *Of Divine Warning: Reading Disaster in the Modern Age*. Boulder and London: Paradigm Publishers, 2009.

Heidegger, Martin. *Being and Time*. Trans. John Macquarrie and Edward Robinson. New York: Harper and Row, 1962.

Howard, Michael. *Liberation or Catastrophe?* London and New York: Continuum, 2007.

Jaspers, Karl. *Tragedy Is Not Enough*. Trans. Harald A.T. Reiche, Harry T. Moore, and Karl W. Deutsch. Boston: Beacon Press, 1952.

Johns, Alessa. "Introduction." In *Dreadful Visitations: Confronting Natural Catastrophe in the Age of Enlightenment*, ed. Alessa Johns, xi–xxv. New York and London: Routledge, 1999.

Keys, David. *Catastrophe: An Investigation into the Origins of the Modern World*. London: Century, 1999.

Mercier-Faivre, Anne-Marie, and Chantal Thomas. "Écrire la catastrophe." In *L'Invention de la catastrophe au XVIIIᵉ siècle: Du châtiment divin au désastre naturel*, eds. Anne-Marie Mercier-Faivre and Chantal Thomas, 7–31. Genève: Droz, 2008.

Nietzsche, Friedrich. *The Birth of Tragedy and the Case of Wagner*. Trans. Walter Kaufmann. New York: Vintage, 1967.

O'Dea, Michael. "Le Mot 'Catastrophe'." In *L'Invention de la catastrophe au XVIIIᵉ siècle: Du châtiment divin au désastre naturel*, eds. Anne-Marie Mercier-Faivre and Chantal Thomas, 35–48. Genève: Droz, 2008.

Sloterdijk, Peter. *Thinker on Stage: Nietzsche's Materialism*. Trans. Jamie Owen Daniel. Minneapolis: University of Minnesota Press, 1989.

Smil, Vaclav. *Global Catastrophes and Trends: The Next 50 Years*. Cambridge, MA: The MIT Press, 2008.

Tolstoy, Leo. *The Kreutzer Sonata*. Trans. Isai Kamen. New York: Modern Library, 1957.

Walter, François. *Catastrophes. Une histoire culturelle, XVIe–XXIe siècle*. Paris: Seuil, 2008.

Isabel Capeloa Gil
(E)Spectating Disaster: A Cultural Condition

Although the observation of disaster has always been a constant of anthropo-logical existence, over the past century it has become a condition of what The-odor Adorno insightfully named modern damaged life (*Minima Moralia*). From Walter Benjamin to Susan Sontag, the spectator of catastrophes has increasing-ly turned into a trope of modern cultural reflection, giving vent to anxiety over the management of the risks and hazards that mark the subject's relation to nature and society. The observation of past catastrophes has been bedeviled by the expectation that some catastrophic event will take place in the future.

The following discussion speaks to the need of understanding the role played by culture in the management of disaster. Arguably, the singularity of any given disastrous event requires the "hermeneutic composability" (Bruner 1991) provided by art and culture to be articulated and this works with the bene-fit of hindsight. Because, as Mary Douglas and Aaron Wildavsky argue (1983, 25), the perception of risk is a social process that depends on a cultural combi-nation of confidence and fear, the acting out of past disasters by the cultural system contributes to fostering resilience and providing for the working through of traumatic contingency and for the mending of damage in the future. Stem-ming from this contention, I shall discuss the relation between the spectatorial position and the cultural expectation of disaster by looking particularly at the performativity of the spectator's role and the management of affect in the aes-thetic representations of disaster in two different case-studies. I am interested in examining the spectatorial position, particularly the way it both partakes of and is distinguished from the role of the witness, the relation of spectating to per-formance and how it mediates between past representation and preparation for the future. The argument proceeds by probing, on the one hand, the *ex-post* performance of disaster in the site-specific Headlong theatre production *Decade* (2011) and, on the other, the *ex-ante* construction of disaster in photographic imagery taken immediately before a catastrophic event occurs. It is my conten-tion, that as these enlightening instances show, spectating, spectacle and antic-ipation are singularly articulated by cultural perceptions and strategies. Turn-ing catastrophe into art is thus not simply an outlet for trauma but a stimulant for resilience in an overwhelmingly uncertain world.

1 Avant propos

It has almost become a truism to declare that the modern subject is a spectator of disaster. Stemming from the Latin etymons *spectare*, to watch, observe, to look at, and *specere*, to look at, to see, spectating brings together cultural intentionality (the will to observe, to watch) and sensory capability (to see). And yet, although spectating supports anthropological identity, the spectator is a misnomer, a universal abstract and an unproductive category. Indeed, I argue that for cultural theory, more than spectators, there are spectatorial positions which articulate distinct subject positions and structures of feeling and are supported and constrained by forms of mediation. As Hermann Kappelhof argues from the standpoint of media and film theory, when it comes to art the spectatorial phenomenon is a "medial vermitteltes, ästhetisches Konstrukt und nicht ein psychologisches Phänomen" (Kappelhof 2004, 18). Hence, while spectatorial positions, as situated constructs, disclose ontological views (contemplative spectator in Lucretius), ethical and moral concerns (moral spectator), compassion and pity, anger and disgust (empathic spectator), political and social judgment (citizen-spectator), aesthetical responses or even libidinal investment (*voyeur*), they are continuously articulated by culturally situated forms of mediation.

My standpoint is thus that of taking the spectator of disaster as a strategic abstraction, articulated by differing spectatorial positions and grounded on mediatized forms of culture. Because the spectator is not a natural category, in the strategic abstraction of the spectator of disaster resonates the imagination of catastrophe and the ambiguous logic that fixes the observation of calamity into both a retrospective and a prospective mode. That is, the observation of disaster prompts looking back to events, fitting them retrospectively into a cogent communal story, whilst allowing a prospective position, whereby a disastrous present becomes a prequel for the imagination of future observations. As Walter Benjamin famously wrote in the *Passagenwerk*, in the age of catastrophe "each era sees the image of its future emerging before its eyes" (1974, 1, 47).

Hence, the sighting of the Hindenburg on the eve of the Lakehurst tragedy on 6 May 1937 (fig. 1) over the Azores islands is an observation of disaster. Despite the absence of clear marks of catastrophe (fire and brimstone, yellow and red), the ship passing by the Azores islands is simultaneously a retrospective observation of disaster and breeds the imagination of prospective airborne disasters. Thus, more than seeing, spectating brings the category of imagination into play. The intentionality of observation is mingled with the expectation of calamity, of a revelation of the accident as *invention*, that which Paul Virilio names "l'accident est une œuvre inconsciente, une *invention* au sens de découvrir ce qui était caché – en attente de se produire au grand jour" (Virilio 2002, 23). Spectating becomes (E)spectating, unavoidably intermingled with the spectacular anticipation of disaster.

Fig. 1: The Hindenburg over Santa Maria (Azores). Unknown photographer

The age of catastrophe, modernity, has indeed created an array of spectatorial positions that enlighten its ambivalent economy. From Diderot's aesthetical ideal spectator (1767) to Freud's voyeur, Walter Benjamin's helpless Angel of History, Janet Staiger's perverse film spectator or Luc Boltanski's passive and frozen subject, a spectatorial culture of disaster has evolved into a marker of modern subjectivity.[1] The awesome spectacularity of tragedy, already an-

---

1 Christian Ruby reflects in his recent *La figure du spectateur* on a culture of spectating in modernity that he describes as the "classical spectator", actively taught to judge by the arts,

nounced by the Greek term *thaumaston*, resonates in the modern observer's wonder before the horrible, famously described by Edmund Burke as a degree of delight in the real misfortunes and pains of others.[2] From the psychoanalytical perspective, the libidinal investment of the spectator in disaster may be conceived as the Freudian slip of an unconscious life yearning for dis-order, while a behavioral scientist may consider this attraction for the observation of the horrible a mere instance of the biological determination of humans to be attracted to what is radically different, strange, alien or frightening. Leaving the biologist assumptions aside, I am interested in the discussion of the cultural place of catastrophic curiosity in the modern structure of feeling. I argue that beyond the libidinal economics of desire, the observation of disaster is a socially constructed cultural practice that has been worked through in social narratives and art forms bridging the gap between the numbing experience of tragedy and its representation.

2 Spectatorial Paradigms and the Perverse Logic of Late Modernity

Greek and Roman Antiquity have provided two spectatorial paradigms that resonate in the logic of modernity: the contemplative spectator and the horrified or Medusa observer. The first is discussed by philosopher Hans Blumenberg in Chapter 3 of his *Shipwreck with Spectator*, devoted precisely to the "Aesthetics and Ethics of the Spectator" (Blumenberg 1996). There, Blumenberg resorts to Lucretius' *De Rerum Natura* (50 BCE) and his description of the spectator of catastrophe, observing a shipwreck from safe, dry land, as a paradigm of existence:

> 'Tis sweet, when, down the mighty main, the winds
> Roll up its waste of waters, from the land
> To watch another's labouring anguish far,
> Not that we joyously delight that man

---

the "modern spectator", a modern participant in the complexity of life and dispersed through manifold spectatorial positions and the contemporary modern art spectator, emancipated from the act of seeing itself and reflecting back on her/himself and, in so doing, reconfiguring art (Ruby 2012, 8–9)

2 See *A Philosophical Enquiry into the Origin of Our Ideas of the Sublime and the Beautiful* (1757): "I am convinced we have a degree of delight and no small one, in the real misfortunes and pains of others" (Burke 1990, 42).

Should thus be smitten, but because 'tis sweet
To mark what evils we ourselves be spared;
'Tis sweet, again, to view the mighty strife
Of armies embattled yonder o'er the plains,
Ourselves no sharers in the peril; but naught
There is more goodly than to hold the high
Serene plateaus, well fortressed by the wise,
Whence thou may'st look below on other men
And see them ev'rywhere wand'ring, all dispersed
In their lone seeking for the road of life (Lucretius 2004, 1–14).

The poem enunciates a contemplative view of the world, a paradigm of distanced observation that was to be that of the philosopher/artist. Seafaring is described as unnatural, a deliberate effort on the part of mankind to supersede the limits put forth by human existence in sharp contrast with the poet's ethics of distance set to avoid danger and keep the limits. In Lucretius' poem the observer's engagement is aesthetical, s/he is literally drawn by sensory perception to reason with the daunting experience of life, while avoiding moral identification: "But naught/There is more goodly than to hold the high serene plateaus, well fortressed by the wise/Whence thou may'st look below on other men/And see them ev'rywhere wand'ring, all disperses/In their lone seeking for the road of life."[3] The contemplative spectator of life, the subject that takes no risks that Lucretius puts forth is sharply contrasted with the second empathic paradigm, the Medusa observer's gorgonic gaze.

The Medusa, as depicted in Caravaggio's famous painting from 1595 has become the icon of traumatic seeing. It has been read as a trope of social and individual trauma, as cunningly suggested by Sigmund Freud, who links the view of the severed head of the Medusa to the fear of castration invoked by decapitation. "To decapitate = to castrate. The terror of Medusa is thus a terror of castration that is linked to the sight of something" (Freud 1990, 47). Consequently "The sight of Medusa's head makes the spectator stiff with terror, turns him into stone" (47). The appalled spectator is thus powerless, stiff with terror, and utter-

---

3 As Blumenberg points out, modernity brings with itself a way of engaging existence as a match against the odds. The Enlightenment for instance views seafaring as a metaphor stressing the danger that must be paid in order to achieve a high return for life: "In complete contrast to this, it will be one of the fundamental ideas of the Enlightenment that shipwreck is the price that must be paid in order to avoid that complete calming of the sea winds that would make all worldly commerce impossible. Through this figure is expressed a justification of the *passiones*, the passions, against which philosophy discriminates: pure reason would mean the absence of winds and the motionlessness of human beings who possess complete presence of mind" (Blumenberg 1996, 29).

ly at the mercy of the daunting monstrous feminine.⁴ The gendering of this traumatized gaze has been abundantly dealt with by feminist and psychoanalytical criticism.⁵ In their varied approaches, from Hélène Cixous to Teresa de Lauretis, Medusa has been the icon of a way of looking that involves risk as Marjorie Garber and Nancy J. Vickers contend (2003, 5). In fact, Medusa's gaze has become the epitome of the disaster spectator's tragic condition, drawn to a seeing that will ultimately cause the seer's demise.

Let us consider for a moment the story of Medusa. Though ancient sources differ on some details, Medusa was one of the three Gorgons of Greek myth and shared the terrifying appearance with her sisters Stheno and Euryale. These monsters with their hair made of snakes were said to possess huge teeth and beards. They were also the owners of a powerful shrieking voice and a gaze that turned all those who beheld them into stone. Medusa (also Médusa, "the Lady"),⁶ unlike her kin, was mortal. This is why Perseus, in order to save the young Andromeda and aided by the goddess Athena, could resort to trickery to approach the monstrous being and capture her head. Cloaked in a cape that rendered him invisible and armed with a mirror, Perseus approached the beast in arrears, guiding his moves with the aid of the mirrored reflection. When the Gorgo's gaze hit the mirror, she saw her own reflection and turned herself to stone. Medusa presents the phobia of the gaze, the horror of perception in a world marked by catastrophe, and frozen to stone by the panic of seeing. As a metaphor of the observer, the monstrous body of the Medusa presents the dislocation of the spectator of catastrophe from the order of things. Unlike Lucretius' contemplative observer, unmoved by pity or despair, Medusa's glaring eyes and guttural cry present in Italian philosopher Adriana Cavarero's words the "human appalled by his very being and contemplating the unspeakable act of his own annihilation" (Cavarero 2005, 25).

In fact, the Medusa-like spectator also avoids interpellation, appalled as s/he is at the horror staring her in the face. Truly, neither the contemplative

---

4 Freud's theory nonetheless does not perceive the Gorgo as an endangered being, who looks onto death herself, but rather sees the character as the horror that wreaks destruction on those beholding her. The gorgonic gaze however draws on the trope of the Medusa as a damage self-inflicting gaze, she is the ultimate spectator contaminated and annihilated by the horror of disaster.

5 From Freud's famous "Medusa Head" (1922) to feminist critique in the 1980s, Medusa has been a recurring trope to address the trials of female identity. See particularly Cixous 1976; de Lauretis 1984, Garber and Vickers 2003.

6 On the duplicity of the Medusa and her origins as a beautiful damsel turned to a monster for repealing the advances of the god Poseidon, see Vernant 1985; Gil 2007.

spectator nor the Medusa observer learn anything. The former because s/he forfeits engagement with risk and does not accept the hazards of life; the latter because s/he is numbed by the horror of existence. Together with Benjamin's Angel of History pushed back by the winds of progress and looking upon a pile of ruins, these spectatorial positions avoid the risks that come with seeing. In a way, Benjamin's dialectical figure shares with the two models a pre-modern gesture, placing the observer of catastrophe as a counter-subjectivity to the modern narrative of instrumental reasoning and progress, based on the wager with risks and odds.

As Kevin Rozario claims, catastrophe is a foundational narrative in the articulation of modernity and its catastrophic logic, negotiated between the understanding of modernity as disaster and anti-disaster, i.e. of modernization as a quest to make the world more secure (modernity as anti-disaster) through development patterns that move through cycles of ruin and renewal, bust and boom, destruction and construction, producing as their collateral damage a myriad of social conflicts as well as technological and environmental hazards (modernity as disaster) (Rozario 2007, 10).

In other words, catastrophe culture is the mark of modernity as a risk society, one that demands the existence of risk to thrive and succeed as claimed by Niklas Luhmann (1988, 38). And yet, though living through risky undertaking, this very model is threatened by the naturalization of risk as a discourse that instead of prompting advancement creates a sort of resilience to the horrible heading towards an apathetic surrender to hazard (140). Furthermore, between these two positions, that of perceiving disaster as a trigger for cultural and social action and the numbing power of devastation, is the spectator disaster constructed.

The catastrophic logic of modernity, Kevin Rozario addresses, draws from an acceptance of the disastrous as a necessary, daunting, albeit stimulating, part of life. Indeed, the untamed irrationality of risk, that which cannot be controlled and is unacceptable for the community, is transformed into a game of rationally-based probability, as the feature that mediates between the utter uncertainty of the outside world be it in nature, in war or in the economy and the cultural determination of societies to function within the daunting prospects of an environment proving increasingly perilous. Thus, while risks are in theory uncontrollable, culture seeks to take control of them by creating insurance narratives formed out of a selection of hazards managed by probabilistic approaches. Modern culture then, whilst allowing human creativity to thrive by gaming with the odds, nonetheless supports a level of acceptable risks within the secure limits of human social existence.

Then again, the observation of disaster, although comfortably wrapped in the cultural cloak for the purposes of this discussion, speaks to other dimensions of the human. More than the experience of fear, famine and fright, the observation of disaster is, as Slavoj Žižek argues in his study of 9/11, libidinally invested (Žižek 2002, 6), marked with a drive for pleasure and death and more often than not already framed by catastrophic experiences broadcast in the media and in this particular case by Hollywood disaster movies. A forerunner to this mediatized indulging in disaster and the ambiguous pleasure deriving from the experience itself is remarkably illustrated, almost one hundred years earlier, in an event narrated by philosopher William James.

On 18 April 1906, the ground shook in San Francisco. The 8.25 earthquake on the Richter scale crippled the water supply and nullified the city's efforts to fight the blazes that would engulf the city for days. Three thousand people were killed and 200,000 were left destitute with 8,000 buildings gutted. William James was visiting Stanford University at the time the events unfolded. The force of the earthquake threw him from the bed, but far from experiencing fear, he saw the experience as "memorable, an encounter with life I wouldn't have missed [...] for anything." In the essay "On Some Mental Effects of the Earthquake" (1906), he described his feelings: "The emotion consisted wholly of glee and admiration; I felt no trace whatever of fear; it was pure delight and welcome" (James 1987, 1215). James describes the life threatening experience as a stimulant for life, an emotion he certainly does not claim solely for himself. Disaster excitement had, in his words, taken Stanford by storm: "Everybody was excited, but the excitement at first, at any rate, seemed to be almost joyous. Here at last was a *real* earthquake after so many years of harmless waggle!" (1221).[7]

Conveying his very own libidinal investment in disaster, James' excitement is undoubtedly representative of the longing for the real that marked so many modern intellectuals, the longing for that radicality of being that could only be tested on the threshold of annihilation. Arguably, he displays the way in which the consciousness arising from the experience of disaster gives vent to a specific form of modern subjectivity and of the modern *habitus*.

The modern spectator is thus simultaneously an observer of the horrible and a participant thereof. Whereas the contemplative spectatorial attitude and the horrified, numbed observer do not seem able to act upon the observation of disaster, the modern attitude suggests a full embedding of spectating, experiencing and acting. Again, as Christian Ruby suggests (2012, 9), the modern

---

7 On the American experience of calamity, see Rozario 2007, 102ff.

spectator of catastrophe is not a universal, abstract category, constructed for aesthetic purposes, but becomes a situated, fragile identity negotiated precisely in and with the observation of calamity. The modern experience of disaster is nonetheless strongly influenced by the nascent reproducible technologies. Film and photography would challenge not only the immediacy of disaster but also inaugurate a new approach to the notion of experience itself. In her illuminating *Regarding the Pain of Others*, Susan Sontag (2003, 18) writes that: "Being a spectator of calamities taking place in another country is a quintessential modern experience". Sontag clearly places the spectator beyond the risky reach of the event and sees the modern spectatorial position as being marked by mediation and distance. Positioned before the technological frontier of the screen or the newspaper page, the observer is seemingly helpless beyond the disgrace of others as the contemplative man on the shore by Lucretius. While technology seems to have brought about the death of distance in social-geographical and temporal terms, suggesting the overwhelming presence of the catastrophic ever present even when happening in the distance, it has also had numbing side-effects. Media and cultural theorists, Luc Boltanski for example, on the one hand, stress the indifference to the suffering occurring in the distance (Boltanski 1993), whilst others, such as Susan Moeller, and on the other hand, view the growth of a usury of compassion arising strangely from a certain compassion fatigue (Moeller 1999).

The problem with some of these theories is the way they emphasize a traditional view of media spectators as homogeneous communities numbed by alienating media effects. Too much mediatized suffering provokes indifference, which, as Annabelle Sreberny writes, seems to have simultaneously become the mark of the modern observer and his/her nemesis (Sreberny 2009, 322). Yet, and because spectating is a situated process involving a negotiation of identity, geography, ethnicity, gender, age, class, the observation of calamity is necessarily articulated from a variety of positions that cut across the wholesome 'we' or 'they' media theorists consider. The variety of spectatorial positions involved in the observation of 9/11,[8] for instance, ranging from the sympathetic Euro-American spectator to some antagonistic Middle Eastern viewers, provides for a systemic framework that speaks against the generalization of apathy or passivity in media viewing. Instead, I would rather prefer to stress the perversity of the spectator, not in the Freudian sense of the word, but in the same vent argued by film theorist Janet Staiger, as a category that escapes the blueprint of expected behavior and activates individual beliefs, traditional narratives and situated

---

8 See Silverstone 2009, 161–170.

emotional states, displaying the irreducible singularity of the human (Staiger 2000, 26).

Fig. 2: Arthur Fellig (Weegee), Their first murder, Oct. 9, 1941. The Naked City (1975, 94) International Centre of Photography/Getty Images

Arthur Fellig's, alias Weegee's, 1941 photograph "Their first murder" (fig. 2) is enlightening in this respect. Considered almost as a paradigmatic figuration of spectatorial perversity, this is a photo about scopophilia and the overwhelming drive to see the apparent demise of others. The photographer was famous for his onsite shots of murder scenes and became known for an uncanny sense of anticipation and the photographic construction of a *noir* urbanity. In 1945, he published some of his most famous compositions in the book *Naked City*. The photograph was captioned with the following lines: "A crowd gathers in the Williamsburg section of Brooklyn, New York City, to see the corpse of Peter Mancuso, shot twice by an unknown gunman as he sat parked at a traffic light. The crying woman is Mancuso's aunt and the little boy tugging the hair of the girl in front of him is her son" (Fellig 1975, 94). The composition displays differing emotional states and indeed provides almost a catalogue of spectatorial positions: the curious, if naïve, observer; the joyfully perverse spectator; the expecting onlooker; the disgusted watcher. All seem trapped by the economy of the gaze in a dangerous pact, trapped, that is, by the invisible logics of the horrible off frame. In the expectant faces of the children watching their first murder we see the scopophiliac drive of the spectator of disaster. The trope of childhood blissful innocence is destroyed by a contentious blend of gorgonic horror and curious perversity, which is a brand of the modern media-made spectatorship.

The modern, fragmented and mediated spectator rises in an age that has often been considered the age of the witness. However, it seems that unlike the witness, who suffers the odds of chance and is marked both by the ill-star of disgrace and the trauma of catastrophe, the spectator observes beyond the economy of disaster, from safe land. The screen that connects is also a border of protection. The cultural category of the spectator seems thus to be clearly distinguished from that of the witness who cannot speak. And yet, the protective shield of ontological distance remains useful to bear witness to what would otherwise be rendered mute by bodily trauma.[9] Witnessing brings responsibility and an ethical commitment to testify to the horror. Between Lucretius' contemplative shipwreck spectator, delighted in the pitiful observation of the disgrace of others but reassured of his security and the logic of Simonides, the poet depicted by Cicero in *De Oratore* as the founder of the art of memory,[10] lies the abyss of testimony. The witness can, while the spectator cannot. In his study on 9/11 audiences, Eric Rothenbuhler interchangeably applies the terms viewer, spectator, witness. Indeed he argues that "on that morning, viewers did not just watch the events unfold. They were also witnesses. The experience conferred upon them ritual power" (Rothenbuhler 2009, 285), building a "persistent feeling of responsibility" (294). Although the blend is useful for Rothenbuhler's argument, the fact of the matter is that the category collapses immediately on being asked to deliver what the witness is required to do, that is, to act by testifying. As the exceptional moment of viewing fades, so does the singularity of

---

9 It is wise to recall the classical Freudian definition of trauma as an experience of heightened intensity that cannot be absorbed and worked through by the psychic apparatus. Trauma provokes a numbing that prevents the subject from acting on the intense memory of violence and articulating the experience through speech.

10 In *De Oratore*, Cicero tells the story of the poet Simonides of Ceos who is reported to have been the founder of the art of memory. Simonides was dining at the house of a rich nobleman and left the house just before the roof collapsed, killing all the dinner guests and the hosts. Because the bodies were so disfigured that they could not be recognized, Simonides was asked to identify them by recalling the place where they had been sitting during the dinner party. Simonides is said to have discovered from this experience that order is what brings most light to our memory (Cicero 2001, 218–219).

the observation, as the spectator once again returns to the logic of ordinary life, while the fire fighters, the policemen, the survivors from the towers will be forever trapped in the systemic economy of the catastrophic event. Bluntly put, while witnessing may involve spectating, the contrary was rarely true up until the scale of violence of the twentieth century blurred the boundaries. The witnesses of the liberation of the concentration camps, such as Ohrdruf, albeit articulating different ethical positions, are set to bridge the gap between observation and action, risking testimony.[11]

Arguably, the spectator of catastrophe is a different survivor in the logic of representation, holding a position as mediator in the after-life of disaster, which often overlaps but is not truly identical with witness status. As Hoffmann and Oliver-Smith argue "Disasters don't simply happen" (Hoffmann and Oliver-Smith 2002, 3). Disasters happen but only become culturally productive when integrated into a system of belief and into a tradition that will frame them within the life-stories of the community. The spectator is precisely the enabler in the after-life of disaster, the one who makes disaster mean when the witness is at a loss of words. The spectator is a cultural enabler but does not necessarily share in the ethical responsibility of the witness to testify to disgrace.

The controversy surrounding the picture of Fabienne Cherisma, a victim of the 2010 Haiti Earthquake, replicates the contentious mingling of spectating and witnessing. The photograph of a 15-year-old girl, shot dead by Haitian police for shoplifting, taken in Haiti in 2010 by Swedish photographer Paul Hensen, won the 2010 Swedish Photo Prize and provoked commotion when it was published in newspapers across Europe. The grayish background of a smoky sky frames the shot of a cement platform over which the body of the dead Fabienne rests. Dressed in a pink frock and sweater, the striking color of the clothes matches the line of blood that drains from her forehead. In the background, oblivious to the tragedy, two looters walk, carrying their heavy burdens. The photo has almost mystical overtones, with the low-angle shot and the position of the body, elevating the scene to a mythical ritual scenery of European religious painting.[12] The lucky enframing of Hensen's photo iconically displays the meaninglessness of life in the exceptional state of calamity whilst unavoidably aestheticizing the innocent beauty of a dead child. However, when a few months later Nathan Weber's reverse shot of the event surfaced, the out-

---

11 Barbie Zelizer (2000) identifies four distinct witnessing positions in the aftermath of liberation: the victim, the journalist, the allied liberator, and, on the other side, the perpetrator.
12 See, for instance, Laurent de la Hire, "Abraham's Sacrifice", 1650, Musée de Beaux-Arts de Reims.

cry over the hypocrisy of journalism and the hoax of empathy spread. Weber's reverse shot depicts a group of photojournalists almost bending over the body to snap the perfect picture. The hallowed illusion of Hensen's picture is abruptly smashed. The journalist, clearly a witness tasked to testify to barbarism, is drawn by a mandate to show and thus to entice a viewer who by being moved will recount the spectating/witnessing of the barbaric act. Notwithstanding the dangerously aestheticizing enframing, Hensen's picture is a beautiful, lucky shot that works. And a picture, even in photojournalism, works as a representational device whose success is measured by the degree of impact it causes on its viewers. It works, thus, as both testimony and representation. Nathan Weber's picture on the other side reports on the task of the journalist and its denunciatory qualities are rather more in the eye of the beholder than on the photograph's intentionality. It deconstructs the (un)intentional fallacy of iconicity and conveys the ultimate modern experience of spectating. That is, it speaks to the fact that the mediated viewing of our late modern world is complex and inhabited by multiple, disparate and even antagonistic viewing positions that often contradict, overlap, comment on, subvert but also engage with and explain each other. Clearly, as Susan Sontag argues, photographers, these star witnesses, do not hold power over their work. They simply convey a position that will unavoidably collide with the viewer. In fact. "The photographer's intentions do not determine the meaning of the photograph, which will have its own career, blown by the whims of the diverse communities that have uses for it" (Sontag 2003, 39). Ultimately, denunciation lies in the informed looking of the media spectator.

The ethical distance between the spectator and the witness and the sometimes strategic interchangeability of the two terms acquires a critical distancing in the theatrical construction of the affective spectator. The third and last paradigm under consideration, that of the affective spectator, is suggestive of the post-critical logic of late modernity and is tested on the basis of the latest production by the London-based theatre company Headlong and its site-specific experiment *Decade*.

The affective spectator
Location: Commodity Quay, St. Katherine's Dock, London, England.
Date: 1 September 2011.
Event: *Decade*, a site-specific theatre production by the London-based Headlong Theatre Company.
Director: Rupert Goold

Two towers. Ten years. Twenty plays. Thus reads the advert announcing the newest Headlong theatre production. The occasion is the anniversary of 9/11 but so is the denouncing of the security state, the governmentalization of daily life and the pleasure in the pay-per-view demise of others. On entering the warehouse where this site-specific production takes place, audience members are drawn into the logics of the security state and to a border crossing between the securitized reality of London on 1 September 2011 and the counterfactual state of exception of theatre. The border space of the lobby in this former trading hall is embellished by official posters of the Office of Homeland Security, the public herded into the winding security lines that lead up to metal detectors controlled by a border officer. After going through security, audience members are questioned by actors in Homeland security uniforms. Where do you come from? Are you a resident? What is your business? And also, who have you voted for in the last election? In this ad lib acting the gap between the stage and the audience collapses as the spectator-actor enters *Decade*'s exceptional performative space. This is simultaneously a space of exception distinguished from the reality of the outside world and a political space of exception where the normative rules of democracy are threatened by what Nestor Garcia Canclini has called the oblique powers that rule the social (Canclini 1990, 25). The double exception of *Decade* is both theatrical and political. But let us, for now, leave the interpretative mode to follow the spectator's immersive plunge into the theatrical plot.

After having been granted rightful admission to the stage by the theatre's border police, the spectator descends into a dreamlike area: a remake of the Windows on the World Restaurant atop the WTC's North Tower. S/he is then led by a cast member to one of the reserved tables where drinks and some food may be ordered until the play within the Headlong play actually begins. The ambiance is cheerful, with a ring of the New York feel prompted by moody Sinatra songs to suggest a relaxed and cosmopolitan environment. Conversations are dim, polite. The waitresses helpful and engaging. The long windows/screens to each side of the room provide a breathtaking view of the Big Apple. Everything is set for a pleasant night or ... for a huge catastrophe.

And suddenly it begins, in this remake of a perfect blue sky day, a whizzing sound approaches, the lights go off, a big bang. In utter darkness, the room seems to shake. The roaring sound of explosions are mixed with human screams. No image, only sound. The effect: nerve wrecking. The upper part of the stage is lit and on a wide glassed wall the actors gather, gazing into the horizon, speechless, numbed, appalled, hopeless, trapped. Lights off. A sound is heard of something popping and splashing, falling. The show goes on, now with a sequence of sketches, a collage of twenty plays by different writers, from the post-modern maverick Mike Bartlett to Mona Mansour or historian Simon

Schama. While different perspectives and fine threads weave the complex web of 9/11 memory, the fact of the matter is that this multi-writer, multi-modal production is basically about 2011 and the role of theatre as an experience seeking to come to terms with trauma. This is precisely the point made by Headlong's creative directors Rupert Goold and Robert Icke:

> There will be people who rehearse the argument that theatre has no valid role in examining recent trauma and disaster, but, the fact of the matter is that this multi-writer, multi-modal production is basically about 2011 and the role of theatre as agent and essential when asking questions about the things that most frighten and disturb us (Goold and Icke 2011, vii).

Then again, this "immersive theatre experience" (http://www.decadeheadlong.com/about) does not claim an enlightening or critical status. It is not about interpreting 9/11, it is not about providing a meaning, or several meanings. The play has no "claims to neatness" (Goold and Icke 2011, vii), where different voices, political viewpoints and forms engage each other in a classical theatrical *agon*. What is more, it has no historical claim. A monologue by Simon Schama, titled "Epic", brings the experience of memory to bear against that of history:

> Sometimes, memory has it all over History for what actually matters. [...] I can't get the distance you see. Just don't have it. I put some space between now and then, a decade-shape-block of months and years. But it catches up in bad-dream time, invades my body space. I can still smell the bad breath of disaster on my face (Schama 2011, 200–201).

Indeed, it is not about understanding, it is about feeling and about asking questions through the affective mode of art.[13] In the particular case of the high intensity event of 9/11, the theatrical spectacle provides affective conviction to the non-representable, to the non-cognitive. In a world overflowing with meanings, the turn to affect has brought together unlikely bedfellows such as behavioral scientists Silvan Tomkins and Paul Ekman and poststructuralist theorist Brian Massumi (Massumi 2002) within the framework of considering human action beyond the intentionality paradigm.[14] Although the ongoing discussion on the possibilities and aporias of affect for cultural theory is beyond the scope of this analysis and certainly deserving of deeper inspection, affect is a useful tool to address *Decade's* immersive site-specific theatrical experiment as an outlet to

---

13 On art and affect, see Clough 2007; Massumi 2002 and for a critique, Leys 2011.

14 According to Ekman, affective processes are independent of intentionality or cognition. They take place in the body before cognition and are utterly pre-cognitive. They occur as well prior to the work of culture or ideology, that is, before meanings, reasons and beliefs. See Leys 2011, 437.

consider both the position of the subject of catastrophe and that of the subject as spectator.

In fact, the immersive experience dwells on shocking the senses and building on the appeal of this multi perceptual shock (haptic, visual, olfactive, auditive and gustatory) to provide a renewed form of aisthetic experience.[15] The theatre becomes a counterfactual, spectacular space where the intensity of the catastrophic event is appropriated by means of affect. The spectator of this modern catastrophe is moved to feel but neither by the events nor by the characters. And here lies the difference between this affective mode and that of the Aristotelian theory of tragedy whereby the spectator is drawn to identify with the misfortune befalling the characters and hence to feel pity for their fate and terror at the prospect of going through the same tragic mistake. The compassionate identification between spectator and the events on stage, which supports tragic catharsis, is alien to the immersive experience of affect conceived by *Decade*'s production, whereby the spectator is sensorially interpellated to feel but not to identify emotionally. Brian Massumi's insightful distinction between emotion and affect in *Parables for the Virtual* is enlightening in this matter. He distinguishes between affect and emotion: "Affect is most often used loosely as a synonym for emotion. But [...] emotion and affect – if affect is intensity – follow different logics and pertain to different orders" (Massumi 2002, 27). The theorist goes on to argue that:

> An emotion is a subjective content, the sociolinguistic fixing of the quality of an experience which is from that point onward defined as personal. Emotion is qualified intensity, the conventional, consensual point of insertion of intensity into semantically and semiotically formed progressions, into narrativizable action-reaction circuits, into function and meaning. It is intensity owned and recognized (28). [...] the affective is not about empathy or emotive identification, or any form of identification for that matter (40).

In plain words, emotion is causal and framed by cultural predispositions. It calls for meaning and is meaningful, while affect rejects causality, intentionality, identification. It is pure feeling. No wonder that after Massumi's lecture at Cornell's School of Criticism and Theory, an audience member is quoted as having said: "If you don't understand, try to feel. According to Massumi it works" (Leys 2011, 434).

The affective spectator may be an intentional fallacy, just as the distinction enunciated above between emotion and affect is also not completely resolved.

---

15 I am referring here to an aesthesis as that which involves sensory perception and understanding without cognition through the senses.

Yet, the immersive performance that is heir to the tradition of post-dramatic theatre (see Lehmann 1999), a theatre that forfeits intentionality, that progresses by means of contraries, bringing together disparate, even opposing views of the almost surreal event of 9/11, suggesting meaning to disturb it a while later, this complex web of feeling that is *Decade* invites questions that provide a glimpse into the not so ordinary spectator of catastrophes of our day and age. More than a counterfactual that acts where referentiality fails to provide a solid structure of feeling,[16] more than a commemoration of the event, or a fiction of representation, the theatrical experience *Decade* provides raises the question of the subject's position before a screen-world of catastrophe. Does the affective spectator who feels disaster before cognition reflect the structure of feeling of the early twenty-first century? How does this position interact with a trope of modernity from Edmund Burke to Walter Benjamin, Maurice Blanchot and Susan Sontag, that of the modern subject as a witness of disaster? What are the ethical concerns of a theory of the subject as aisthetical observer? And finally, how does it impact on communal life in the aftermath of a catastrophic event?

While geo-sociological distinctions between catastrophe and disaster differ on an empirical basis (number of casualties, impact of the event on human communities, degree of devastation, lasting effects), from a cultural-symbolic perspective they are utterly irrelevant. In cultural terms, the degree and impact of disasters are significant insofar as they affect the structure of feeling of a given community, the mode of recollection, the relation with the past and the perception of the future. Said otherwise, disasters are culturally significant because they mean, either by being integrated into the hermeneutic nexus of the community's narratives and discursive practices or precisely because they are unpredictable, flogging expectations and thus requiring different modes of hermeneutic composability to be made to mean. Those who impart this hermeneutical task on the catastrophic are the multiple spectators and witnesses, traumatized or numbed, distanced and illuminated, affective or critical. If disaster is a cultural condition of the modern age that is so because the spectator has become the trope of our ever changing and complex late modern subjectivity.

---

16 Read again Simon Schama's monologue on the role of art and disaster: "It needed to look *fake*, you see, before people could read it as *real*, or as everyone likes to say now, as if real isn't enough, not for extra strength, added-value calamity ... SURREAL" (Schama 2011, 200).

References

Benjamin, Walter. *Gesammelte Schriften V*, eds. Rolf Tiedemann and Hermann Schweppenhäuser. Frankfurt am Main: Suhrkamp, 1974.

Blumenberg, Hans. *Shipwreck with Spectator. Paradigm of a Metaphor of Existence*. Boston: MIT Press, 1996.

Boltanski, Luc. *La souffrance à distance: morale humanitaire, médias et politique*. Paris 1993: Métailié.

Bruner, Jerome. "The Narrative Construction of Reality." *Critical Inquiry* 18, no.1 (1991): 1–21.

Burke, Edmund. *A Philosophical Enquiry into the Origin of Our Ideas of the Sublime and the Beautiful*, ed. Adam Philips. Oxford: Oxford University Press, 1990.

Canclini, Nestor Garcia. *Culturas híbridas. Estratégias para entrar y salir de la modernidad*. Mexico City: Grijalbo, 1990.

Cavarero, Adriana. *For More than One Voice. Toward a Philosophy of Vocal Expression*. Stanford: Stanford University Press, 2005.

Cicero, Marcus T. *De Oratore*. Trans. John M. May. New York and Oxford: Oxford University Press, 2001.

Cixous, Hélène. "The Laugh of the Medusa." *Signs* 1, no. 4 (1976).

Clough, Patricia T., ed. *The Affective Turn: Theorizing the Social*. Durham: Duke University Press, 2007.

Douglas, Mary, and Aaron Wildavsky. *Risk and Culture. An Essay on the Selection of Technological and Environmental Dangers*. Berkeley: University of California Press, 1983.

Fellig, Arthur. *Naked City*. New York: Da Capo Press, 1975.

Freud, Sigmund. *Gesammelte Werke*. Vol. 17. Stuttgart: Fischer, 1990.

Garber, Marjorie, and Nancy Vickers, eds. *The Medusa Reader*, New York: Routledge, 2003.

Gil, Isabel Capeloa. *Mitografias. Figurações de Antígona, Cassandra e Media no Drama de Expressão Alemã do Século XX*. Lisboa: Imprensa Nacional Casa da Moeda, 2007.

Goold, Rupert, and Robert Icke, orgs. *Decade. Two Towers. Ten Years. Twenty Plays*. London: Nick Hern Books, 2011.

Hoffman, Susanna M., and Anthony Oliver-Smith. *Catastrophe & Culture: The Anthropology of Disaster*. New Mexico and Oxford: School for Advanced Research Press, 2002.

James, William. *Writings 1902–1910*. New York: The Viking Press, 1987.

Kappelhoff, Hermann. *Matrix der Großen Gefühle. Das Kino, das Melodrama und das Theater der Empfindsamkeit*. Berlin: Vorwerk, 2004.

de Lauretis, Teresa. *Alice Doesn't: Feminism, Semiotics, Cinema*. Bloomington: Indiana University Press, 1984.

Lehmann, Hans-Thies. *Postdramatisches Theater*. Frankfurt am Main: Verlag der Autoren, 1999.

Leys, Ruth. "The Turn to Affect." *Critical Inquiry* 37, no. 3 (2011): 434–472.

Lucretius, Titus. *On the Nature of Things*. New York: Dover Classics, 2004.

Luhmann, Niklas. *Soziologie des Risikos*. Frankfurt am Main: Suhrkamp, 1988.

Massumi, Brian. *Parables for the Virtual: Movement, Affect, Sensation*. Durham: Duke University Press, 2002.

Moeller, Susan. *Compassion fatigue: How the Media Sell Disease, Famine, War and Death*. New York: Routledge, 1999.

Rothenbuhler, Eric. "Fractura simbólica e processo de reparação: as testemunhas do 11 de Setembro." In *O Terror Espectáculo: Terrorismo e Televisão*, ed. Daniel Dayan, 281–296. Lisboa: Edições 70, 2009.

Rozario, Kevin. *The Culture of Calamity: Disaster and the Making of Modern America*. Chicago: The University of Chicago Press, 2007.

Ruby, Christian. *La figure du spectateurs. Eléments d'histoire culturelle européenne*. Paris: Armand Colin, 2012.

Schama, Simon. "Epic." In *Decade. Two Towers. Ten Years. Twenty Plays*, 197–208. London: Nick Hern Books, 2011.

Silverstone, Roger. "A mediatização da catastrophe: o 11 de Setembro e a crise do Outro." In *O Terror Espectáculo: Terrorismo e Televisão*, ed. Daniel Dayan, 161–172. Lisboa: Edições 70, 2009.

Sontag, Susan. *Regarding the Pain of Others*. New York: Picador, 2003.

Sreberny, Annabelle. "Por quem nos tomamos? O distanciamento jornalístico e o problema o pronome." In *O Terror Espectáculo: Terrorismo e Televisão*, ed. Daniel Dayan, 321–346. Lisboa: Edições 70, 2009.

Staiger, Janet. *Perverse Spectators. The Practice of Film Reception*. New York: New York University Press, 2000.

Vernant, Jean-Pierre. "Die religiöse Erfahrung der Andersheit: das Gorgogesicht." In *Faszination des Mythos. Studien zu antiken und modernen Interpretationen*, ed. Renate Schlesier, 399–420. Basel: Stroemfeld/Roter Stern, 1985.

Virilio, Paul. *Ce qui arrive …* Paris: Actes Sud, 2002.

Zelizer, Barbie. *Remembering to Forget. The Holocaust Before the Camera's Eye*. Chicago: Chicago University Press, 2000.

Žižek, Slavoj. *Welcome to the Desert of the Real*. London: Verso, 2002.

José Manuel Mendes
Catastrophes, the Imaginary and Citizenship: The Production of the Other and the Singularity of Experience

1 Introduction

This article reflects on the relationship between the media and the construction of publics, taking these to be based on identities and the processes of creating identities activated in concrete, well-defined contexts. This relationship, which is complex and defined by power dynamics, allows spaces, processes of citizenship, and the visibility and invisibility of causes, projects and trajectories to be defined. The central question which serves as the starting point for these reflections is the following: what contribution does the media, with its autonomous operational logic, make towards reflections on citizenship? Who and what does it include and exclude?

I intend to approach the specter of theoretical positions on the role played by the media in the production of citizenship from the basis of the polysemic notion of public sphere. Are we demanding too much of the media? Does it really set the agenda for what is relevant in a national and international context? As an opening statement, Laurent Thévenot's comparative analysis of the construction of the European sphere (1999, 77) demonstrates how there is no consistent and sustained civic vocabulary applicable as a discursive resource to counter the current hegemony of the language of the market. Without this vocabulary, we cannot conceive of alternative forms of civic action and interaction.

The first part of this article presents a discussion on the concept of imperfect citizenship and the notion of public sphere in Habermas. The second part approaches Oskar Negt's theorization of the concept of oppositional public sphere as an alternative vision. The third part of the paper deals with the relationship between citizenship, public sphere and risk. The last section considers the media coverage and socio-political impacts of the tsunami in the Indian Ocean in 2004 and Hurricane Katrina in the United States in August 2005 as illustrations of the diffractive complexity of media production and the persistence of colonial discourses. The two events had direct consequences for public opinion in the West and made explicit the discriminatory and socially exclusive criteria contained in a biopolitics of populations based on racism, class difference and discourses that produce fear and dangerous alterities.

2 Imperfect Citizenship and the Community of Fate

The concept of citizenship used in this text aims to range beyond legal and political definitions and, based on the work of Étienne Balibar (2001), employs the concepts of the community of fate and imperfect citizenship. The community of fate implies political conditions in which uncertainty and conflict, if not violence, prevail (2001, 209). In terms of its territorial component, the community of fate may extend from a building, street or neighborhood to the entire planet. As a consequence, citizenship is a complex notion that is always defined and constructed on various levels, within multiple frameworks linked in diverse ways. Imperfect citizenship is therefore constituted from practices and processes rather than being stable or pre-defined.

The synthesis of these concepts of the community of fate and imperfect citizenship enables us to conceive of the public sphere as being pervaded by the weight and contradictions of individual and collective subjectivities. To quote Balibar, "the *public sphere*, in reality, is nothing more than an objectified and universalized representation, a collective control, a communication field for the 'passions and interests' of society" (2001, 212).

This reflection by Balibar, and its attention to questions of identity, derives from his theoretical proposition concerning the notion of the political. For this author, the phenomenon of politics is defined by three concepts: emancipation, transformation and civility (Balibar 1997). The autonomy of politics corresponds to the ethical aspect of emancipation. The possibility of conceiving of transformation derives from the structural and conjunctural conditions of the political and refers to the heteronomy of politics. Civility corresponds to the heteronomy of the heteronomy or, in other words, the contradictions and ambiguities of identities, belonging and rupture.

On the basis of this ductile notion of citizenship anchored in contradictory and complex political processes in which the challenge of identities and identity politics becomes pressing, the traditional notion of the constructed public sphere may be interrogated on the basis of the propositions of Jürgen Habermas. In his pioneering work on the emergence of the bourgeois public sphere in Western societies, this author drew attention to the return of charisma within the representational public sphere, based on image and opinion specialists who convey the interests of economic and political powers (Habermas 1989). This pessimistic vision of Habermas, which was directly associated with the period in which the book was written, would later be transformed into a relative naturalism in his work *Between Facts and Norms* (1996).

As Deborah Cook notes (2001, 139), Habermas seems to assume that the systemic-paternalistic functioning of liberal democracies must be the standard to adopt. Therefore, according to this writer, it is only in rare cases that the public sphere fulfils the conditions that enable it to become a communicative power capable of directly influencing the political system.[1] Moreover, what Habermas calls "civil privatism" is reinforced by colonization of the life-world, making it difficult for a robust and vibrant public sphere to develop. What is surprising about Habermas' new arguments is that the weakness of the public sphere derives almost exclusively from the negative dynamics of the life-world and not from economic or political systems and their hegemonic operational logic.

This new Habermas theoretical proposition ascribes the capacity to make demands on the political system to the actions of social movements, above all through spectacular and disruptive initiatives, given that the media does not fulfill its role in mediating and shaping citizens by drawing on expert opinions.

What possibilities are there for the construction of active and participative citizens? Rather than speaking of a public sphere, which acquires almost metaphysical connotations, it seems more appropriate to speak of publics in an attempt to restore the complexity of media construction and reception and its political impacts. According to Cefaï and Pasquier (2003), publics are not essential, pre-existing entities but are instead produced through performances that publicize a social problem, measure or public policy, or a project, programme, spectacle, etc. The authors distinguish between the media public and the political public. The media public is, above all, the target for a representational mechanism. The political public, in John Dewey's original sense, is "an associative, inquiring or deliberative public which aims to control the consequences of an event or an action and seeks to define the forms of public good" (1991, 18).

This pragmatic approach to publics highlights the importance of social experiences and interpersonal links in shaping personal and collective experiences of reception of the media involving social, civic and political commitment. The forms of collective mobilization reveal the role of the media as the practical operator of experiences of identity and identity narratives. If a political public can be interpolated, reinforced or challenged by the actions of the media, this can also produce public problems and create publics susceptible to conversion into political publics.

---

1 For a decisive analysis of the naturalisation of the concept of civil society and the relationship between political culture and the public sphere, see Somers (1995). For an attempt to recover the concept of civil society applied to situations involving cultural and historical diversity, see Chambers and Kymlicka (2002).

3 The Creation of an Oppositional Public Sphere

As a direct reaction to Habermas's normative and integrative vision in his account and enshrinement of the bourgeois public sphere, Oskar Negt, in proposing a proletarian or oppositional public space, as he would later term it, sought to make visible the collective and alternative forms that give public expression to those human needs that evade the straitjacket of the dominant media representations (Negt 2007, 216). This overspill process allows a public sphere to be constructed that accommodates immediate, lived experience and enables an inalienable democratic order to be founded.

Oskar Negt proposes the restoration of rhetoric and the art of persuasive discourse to disseminate demonstration and protest. In addition, whereas for Habermas public space, free discourse and the affirmation of freedom of opinion in a profoundly European sense represent the factors that legitimize the new bourgeois social order, strictly demarcated in terms of economic interests and the pressures of the state apparatus, Negt configures a broad notion of public sphere necessarily including production, economics and the private sphere.

The potential politics of the production and private spheres must be worked on in order to mobilize them in the service of an enlarged public sphere. Polemical issues originating from production, work and the private sphere and traditionally excluded from the bourgeois public sphere are currently undergoing radical changes (Negt 2007, 222). As Alexander Neumann aptly notes (2007, 8), the Habermas concept of publicity is deliberately presented as the theoretical idealization of a form of politics based on recourse to violence and the exclusion of the majority of the population. The oppositional public space, which flows and consolidates itself in rebellious subjectivities and partial connections, restores, makes visible and projects the particularities and combinations of specific real-life situations. The aim is to start from fragmentary experiences, almost inaudible, unheard murmurs and speech that takes risks (12).

For Oskar Negt, the experiences that overflow from concepts and negate them are more important to a critique than the words used to positively name things. Rebellious subjectivity contains the obstinacy, authenticity and negativity of actors who resist the triumphal march of the winners. Whereas the bourgeois public space proceeds by generalized abstracts, the oppositional public space is directed towards the accumulation of unique experiences. In opposition to normative discourses, oppositional space is characterized by direct speech, which allows for the exchange of experiences and resolution by dialogue that is partial and always open to conflict (Neumann 2007, 21).

4 Citizenship, Public Sphere and Risk

The advent of modernity led to the concept of *chance* being replaced by the concept of risk, expanding the context of trust in addition to the role of the state in regulating this area (Giddens 1991; Luhmann 1993). However, as Boaventura de Sousa Santos has argued (1995), we are also witnessing a widening gap between the capacity to act and the capacity to predict, thus dramatically increasing risks, both in terms of scale and frequency. This presents new challenges for risk regulation by national states, due above all to transnational dynamics and the increasing needs of citizens in terms of security and the existence of well-defined prevention and mitigation plans with clear objectives. As a result, the need arises to study risk regulation systems (Hood, Rothstein and Baldwin 2001) and how they adapt to the need for transnational cooperation, given the globalization of risks. According to Ulrick Beck (2001), "States must *de*-nationalize and *trans*-nationalize themselves for the sake of their own national interest, that is, relinquish sovereignty, in order, in a globalized world, to deal with their national problems." This, for Beck, is an inevitable trend within what he terms the world risk society (1999; 1992).

This separation of nation and state proposed by Ulrich Beck diverts attention away from the founding material and symbolic mechanisms of internal political arenas and political struggles occurring as a result of external events (Klinenberg 2002). This is why the creation of a political arena demands explicit inclusion, the justification and articulation of individual choices and facts and debates made perceptible to specific audiences (Barry 2001; Callon and Rabeharisoa 2004).

Permanent dangerous events and situations tend to increase the legitimacy of state intervention, often involving the suspension of social and economic norms, creating a state of exception indicating the inevitable presence of the state. These extreme events present concrete challenges to the production of knowledge within the social and natural sciences that, in the majority of cases, restrict the supposed autonomy of the sciences.

However, many scholars working in this area reinforce the role of experts and the established powers (states and international agencies) in the management of uncertainty, marginalizing the knowledge and participation of citizens. This clearly reflects in the "enlightened catastrophism" of Jean-Pierre Dupuy (2005; 2002) or the worst case scenarios of Lee Clarke (2005).

This theorization of "extremes" is based on an ecology of fear (Davis 1998) and a politics and culture of fear (Furedi 2005; 2002) which do not allow for any progressive and gradual composition of a shared world (Latour 2005). Therefore, it does not prove possible to create this shared, heterogeneous, complex and conflictual world through a sociology of virulence (van Loon 2002), but instead through a modest, interlinked approach alert to the components emerging out of the social sphere and able to activate and democratically weave a version of this sphere.

Extreme events such as disasters and catastrophes show the political work involved in positioning disposable groups and individuals outside social networks and communities within the national imaginary. The social sciences, as technologies of humility (Jasanoff 2003), must create visibility for individuals and groups located on the interstices of socio-technical alignments and must be involved in the collective work required to describe and personify these individuals and groups within a political process of full citizenship.

The eruption of fear in the arena for public debate, created whether by natural or social events with major material, symbolic and political visibility such as 11 September 2001, the Asian tsunami or Hurricane Katrina in New Orleans, and its explicit deployment within a logic of political control consubstantiated in the politics of fear, demands reflection and the advancement of alternatives for the construction of a shared, conflictual, diverse, heterogeneous world that is, nevertheless, always defined by shared experiences and identities.

These alternative visions, leading to a democracy based on dialogue in which all the interested agents associated with a particular matter have the right to speak and to investigate, are based on constant attention to emerging identities and the creation of alternative scenarios that can respond to local non-equivalent issues in a precarious balance between general and particular interests.

As Niklas Luhmann aptly points out, the hiatus has been growing on a communicational level and in the wake of the logic for the analysis of systems within modern society (namely politics, law, economics and science) between those involved in decision-making and those excluded from this process but nevertheless suffering the consequences of the decisions that are taken. This increasing communicational hiatus may lead to a lower level of trust in relations between members of a given society.

The alternative is to define social technologies for participation leading to the construction of civic epistemologies that allow for the informed and critical presence of citizens within the public sphere (Jasanoff 2005). These civic epistemologies define how democratic societies acquire shared knowledge for the purposes of collective action shaped by different political cultures and national contexts.

The consolidation of the importance of civic epistemologies must be based on experience accumulated through analysis of disaster situations or catastrophes in which, contrary to the affirmations of many specialists and political leaders, the responses of the populations, groups and individuals affected are not based on irrationality or panic (Clarke 2002) but on mutual help, solidarity and the construction of disaster communities shaping the capacity to resist, recover and slowly re-establish bonds, lived experiences and inhabited spaces.

The myth of panic allows political and administrative leaders to retain information that reinforces the logic and dynamics of power in the management of extreme events. Panic, or the presumption that this will determine the actions of individuals, is always ascribed to ordinary citizens and never to leaders, scientists or technicians. This political definition of panic removes the possibility of incorporating citizens as a genuine resource in planning and in the response to disasters or extreme events. Moreover, the failure to divulge information leads to a mistrust of political agents, preventing any appropriate usage of shared knowledge and consolidating the potential for resilience and resistance on the part of populations, groups and individuals.

Rather than an ecology of fear and a politics of fear, the possibility of constructing a public sphere with full citizen participation in relation to hazards, risk and vulnerability implies an alternative vision based on the security of populations (Lakoff 2006). This new paradigm raises a central question: what type of techniques, instruments and government institutions are most relevant in terms of focusing on the welfare of citizens and which areas of knowledge and types of intervention need to be defined to preserve the security of populations?

The framework for analysis must be long-term taking into account provisions for public health and the reduction of poverty rather than responses and actions defined by urgency, the short-term and the mitigation and limitation of damage. Special attention should be paid to structural factors and their spatial dimensions, which require planning and a clear and well-defined context for the actions of public bodies. The definition of sustainable lifestyles necessarily implies sustained and integrated intervention in terms of the welfare of populations.

A logic of civic participation associated with the construction of a public sphere based on dialogue in relation to risk issues must consider the living conditions of human beings as members of a social collective and their right to integration and full citizenship.

5 Extreme Events, Catastrophes and the Racialization of the Exploited

Taking as a reference point local theories on the media, publics and the construction of citizenship discussed above, mainly within a European and North American context, the aim of this second section is to analyze the media coverage of the tsunami that ravaged the Indian Ocean in 2004 and Hurricane Katrina, which destroyed New Orleans in 2005. I follow Walter Mignolo closely (2005a, 405) in his appeal to suspend the theory of proletarianization and concepts of people, popular and the multitude until there is a clear understanding, both in epistemological and political terms, of what racialization means and how it operates in a capitalist and imperial/colonial world. For Mignolo, "a topology of social actors hand in hand with a topology of being is an imperative to understand the current imperial articulation both at the level of the imperial differences and the colonial differences" (2005a: 405).

This proposition may be integrated with the critical Boaventura Sousa Santos view of post-colonialism (2002, 16). For this author, although the colonial relationship is one of the fundamental unequal power relationships within capitalism, it is not unique and should be interlinked to other power relationships such as exploitation based on class, sexism and racism. Analysis of economic factors and the historical and heterogeneous configurations of the colonial relationship should also be considered as an essential component in the analysis of culture or discourse.

The historical geography of disasters reveals evidence of the hegemonic and performative production strategies of an insecure world. As Greg Bankoff clearly demonstrates (2004), at the end of the twentieth century the *topos* of natural disasters replaced the *topoi* of tropicality (disease is resolved by Western medicine) and development (poverty is resolved by investment and Western aid) to produce another dangerous and distant exoticism. However, some catastrophes in the Western world, more specifically Hurricane Katrina in 2005, revealed how this exoticization, based on racial and class criteria and the biopolitics and geoeconomic management of populations, is also fully functioning in certain geohistorical locations (Mignolo) in the hegemonic center.

The tsunami that hit the Indian Ocean on 26 December 2004 marked the emergence of a global disaster community, due to the number of people affected, the range of countries directly hit, the colossal material damage and, more importantly, the media coverage, amplified by the victim status of thousands of European and North American citizens present in tourist resorts across the region.

After the 2004 tsunami, technical staff representing dozens of NGOs from 55 countries, rushed to Sri Lanka. The lack of coordination and the reactive nature of these organizations served to reveal the absence of any predictive capacity or planning, the non-incorporation of theories and structures for the analysis and reduction of risk and the low importance attributed to accumulated learning associated with mitigation and recovery from disaster situations. The accumulated knowledge associated with disaster management had not been generalized or incorporated into the organizational and logistical structures of the NGOs operating in catastrophe situations (Twigg and Steiner, 2002). Their operational logic was appropriate for the socio-historical, socio-political and socio-geographical contexts of post-colonial Latin America, but not adapted to post-colonial situations in Asia.[2]

For example, one member of a local authority in Sri Lanka complained to a journalist from the French *Nouvel Observateur* magazine that they needed cement and electricity and not brioches or croissants. The presence of the United States fleet off the coast of Sri Lanka also created feelings of hostility, above all in the zone controlled by the Tamil rebels.

In addition, in the Banda Aceh area of Indonesia, aid from the UN and the NGOs was co-opted by the central government which used it as a political weapon against rebels (40% of the population in this region were living below the poverty line before the tsunami).

The logic which supports the work of major international aid systems (NGOs and the World Health Organization) may also restrict or annul the work of local organizations. A major catastrophe, like a battlefield, is a real-world testing ground for health and hygiene matters for the World Health Organization. The danger is that a future humanitarian power and health authority is being created which, supported by collective emotions, strips small countries or territories of their capacity to act (biopolitics in an international context). Within this context, India's refusal to accept international aid during the tsunami crisis demands consideration beyond a simplistic analysis of the receptiveness/unreceptiveness of political regimes and their dictatorial logic or willingness in the face of international aid and world charitable organizations.

In taking a concrete approach to the media coverage of the tsunami in the Indian Ocean, we are confronted by the first world disaster both in terms of its dimensions and the unusual involvement in this type of event of tourists who

---

2 The cases of the cyclone which hit Myanmar (Burma) on 2 May and the earthquake on 12 May 2008 in China clearly demonstrate this.

were citizens of the core countries in the world system (with media reports additionally forgetting the presence of thousands of Asian tourists at the scene).

In the initial moments, the information was based almost exclusively on amateur video footage, democratizing the flow of information and reinforcing the stamp of reality. On the other hand, Western governments, contrary to what might have been expected given the dimensions of the disaster, displayed an inertia and an inability to act and coordinate that was almost pitiable and truly striking. Beyond an inability to mourn citizens who had perished in a faraway tourist paradise, what we witnessed, in the case of Sweden for example, was a lack of recognition and an inability to represent an event of this kind, the absence of an enlightened catastrophism in the face of the omnipotence and self-sufficiency of governments that had great economic resources but were incapable of conceiving of a catastrophe beyond their territorial limits.

In the diffractive interplay of images and discourses that followed the tsunami, the inhabitants of Europe and the tourists in the Indian Ocean region realized the weakness of the state as the global therapist in the imaginary of its citizens (Sloterdijk 2006, 266). Cracks opened up in the crystal palace with its warm, self-complacent endosphere and each of the tourists and the disaster community that consisted of their families, friends and loved ones felt completely unprepared, vulnerable and at the mercy of the elements. The shattering of the crystal palace, the radical change in their comfortable homeostasis as world travelers and the consequent experience of vulnerability made them citizens of the world and simple human beings.

Immersed in the illusion and myth of national containment, the European national structures were not prepared to respond to a world disaster community and public opinion in the European countries involved in the events of December 2004 was critical of the actions of their governments during the crisis. This was confirmed in many sociological studies of disaster situations in which before, during and after disasters, the "general public" believed it was important to trust in the authorities and their respect for citizens (Luhmann 1993).[3]

There were many critical analyses of the media coverage of the tsunami by the core countries. Harsha Walia (2005) contrasted what she termed the discourse of compassion and the ideology of humanitarianism associated with

---

3 Examples in Europe include the case of the oil spill off Galicia, the Madrid bombings and the contrast between the official, civil and media indifference in Portugal to the approximately one thousand deaths from the effects of the 2003 heat wave in comparison with the situation in France, where the same event led to collective awareness and a questioning of the concept of citizenship, following the abandonment of thousands of elderly people to their fate and thousands of unclaimed corpses buried in mass graves.

natural disasters with the neglect of the humanitarian crises in Darfur and Rwanda. According to the author, we were facing a re-writing of colonial consciousness and the logic of Western superiority. In her opinion, international aid and the politics of global compassion, including the strong presence of NGOs, would ultimately defuse the anger of populations affected by the tsunami. The capacity for resistance and recovery would be diluted within a logic of victimization.

Peter Philips (2004), a North American sociologist and the director of *Project Censored*, called the stance of the leading North American media hypocritical, contrasting the excessive attention dedicated to the tsunami with the prolonged silence over the death of 100,000 civilians since the start of the invasion of Iraq.

However, we cannot simply restrict ourselves to accusations. In transforming it into a global disaster, the visibility of the tsunami triggered local, national and international solidarity work, mobilizing resources and people on an unprecedented scale. The role of the media in activating humanitarian measures can neither be ignored nor its role in providing effective support for the populations affected.

In the wake of Luc Boltanski (2000; 1999), we instead have to enquire about the principles of an approach to pity which, abandoning the negative version of compassion, enables a politics of pity to be defined, reconstituting its specifically political, controversial and even conflictual dimensions within a democratic logic of citizenship. How can distant suffering be represented and acquire political connotations that will allow for action, revolt and solidarity?[4]

The central issue involves the selection of the victims in question (who, for example, remembers that in 1970 cyclones and floods hit Bangladesh and killed 300,000 to 500,000 people?) as the politics of pity emerge – and the language here is crude and cruel – from an excess of victims, both on a material level and in terms of their representation in the media.

The re-legitimization of humanitarian work and its re-politicization includes the capacity of ordinary citizens to assimilate political events, to feel them in their everyday lives as part of their lived experience and to experience them as challenging personal and collective dilemmas, transforming anger and emotion in the face of the suffering of others into collective action by social movements independent of party politics and the state.

---

4 Luc Boltanski (2000) defines three topics for suffering: denunciation (associated with anger), sentiment (associated with emotions) and aesthetics.

Hurricane Katrina once again confronted developed countries, or rather the hegemonic economic and political power, with the racial and class scars present in land planning, access to assets, and citizenship status. The rhetoric and concepts in question are crucial to the performance and reproduction of the dynamics of power: at the Nairobi Climate Conference in 2006, various analysts referred to the people uprooted from New Orleans as the first refugees of climate change. This is a dubious attribute, since the concept of refugee implies noncitizen status. The uprooted of New Orleans were North American citizens and staked their claims as such.

After the initial inertia of the North American local, state and federal authorities in dealing with the effects of the hurricane, those affected were systematically relocated to nearby and distant states, experiencing an explicit policy of selectiveness and separation of families and communities. The central issue after the event concerned the reconstruction of public housing and the right to return to the former place of residence.[5] Around 200,000 inhabitants have still not returned to New Orleans, raising the question of whether the conditions will ever be in place for them to return one day.

The people affected by Hurricane Katrina directly felt the effects of a structural and insidious racism that lingers in the United States and defines an entire tradition of selective biopolitics. After the catastrophe, the same people found themselves with threshold citizenship status, uprooted by force and removed from the arena and the public spheres.

In order to define their situation in a democratic state of law, Henri Giroux proposes the notion of disposable groups (2006, 10). The biopolitics of Katrina reinforced the idea that the poor Afro-American populations affected were a burden on the federal state and would have to find solutions by themselves. The individuals who made up these groups were seen as having no stable social ties, citizen's rights, contributive or productive capacities. Their initial classification as refugees was an indication and designation "which suggests that an alien force inhabited the Gulf Coast" proposes Henri Giroux.[6]

What, therefore, is the reference point enabling us to positively assess and classify an approach to the media regarding disasters? The following topics are proposed:
–   avoidance of sentimentality and sensational images and news;

---

5 The organization *Common Ground*, founded in New Orleans after the hurricane in August 2005, played an important role in defending reconstruction and the right to return.
6 The evacuation plans for African-Americans followed a meticulous process of relocating members of the same family to states as far away as Utah, Oklahoma, Texas and even Alaska.

- presentation of local strategies for recovery and the reformulation of social networks and complex power relationships based on racism, sexism, economic domination and biopolitics;
- transmission of the complexity of the situation and the real or potential conflicts existing in the area;
- provision for ordinary voices and their definitions of the situation to be heard;[7]
- attention to practices, performances and discourses that indicate the production of an oppositional public space.

In assuming difference, this type of approach allows for a concept of common humanity to be constructed and taking both the strong and weak points of local civil society into account.

Disasters are occasions which clearly reveal the social dynamics of the societies they affect (social structures, social networks, inequalities and the capacity for resistance) and the quality of the state services, their organization and operational logic. They also test the strength of the ties that bind us all together as human beings and citizens of different countries, raising political and practical moral dilemmas that are complex and perhaps impossible to resolve. However, they may also lead to the conception and materialization of alternatives for building a fairer and more equal world.

6 Conclusion

In conclusion, it may be said that the media are central components of democratic societies since they denounce the contradictions and complexities of subjects and problems that pervade these societies and suggest alternatives to be conceived and experienced as ways of resisting hegemonic biopolitical thinking and practices. Resorting to a notion of imperfect citizenship, rejecting simplistic and catastrophic visions of the media's potential to manipulate, the public sphere may be perceived as composed of multiple media and political publics who are active in certain contexts and mere spectators in other situations. As the examples presented here clearly show, the ideological function of the media – defining ideology from a Marxist perspective – also implies that publics, associations and movements within civil society and the deconstruc-

---

7 A masterful and powerful analysis of Hurricane Katrina can be found in Spike Lee's documentary (2006).

tion and critique of grammars of hegemony will only be possible through the active interpolation of journalistic discourse and the public presentation of alternative denominations and definitions, whether or not associated with alternative means of production and dissemination.

Without demonizing the work of journalists, citizens and groups directly targeted by the discourse of these professionals must demand the right of reply and the right to produce other grammars that explode the colonialism (Mignolo 2005b) of knowledge. The contradictions within the field of journalism will become the catalyst for other voices and the possibility for forging new ties or new definitions of citizenship on the basis of the alterity of political and ideological practices and positionings, producing counter-discourses and counter-publics.[8]

Whilst Habermas appealed to social movements as the only agents capable of questioning the political system, we know that the media format and the framework for their actions define the tone and the impression that all events convey to media publics. The competitive logic within the media and the anxiety of its professionals regarding their public's unfamiliarity with its ductile nature, in addition to its often ephemeral logic, create the opportunity for other vocabularies to be projected and, within a dynamic of conflict, other alternatives to be conceived for the construction of a shared but non-consensual world.

The production of an oppositional public space materializes, as Oskar Negt states, through the accumulation of singular experiences conveyed in the conventional or alternative media and through the production of devices to establish rebellious and diffractive subjectivities released within the confines of colonialism and circulating as possible and alternative spaces in addition to, or despite, local theories manufactured in a myriad of academic worlds.

References

Balibar, Étienne. "Trois concepts de la politique: émancipation, transformation, civilité." In *La crainte des masses. Politique et philosophie avant et après Marx*. Paris: Galilée, 1997.

Balibar, Étienne. *Nous, citoyens d'Europe? Les frontières, l'État, le peuple*. Paris: La Découverte, 2001.

Bankoff, Greg. "The Historical Geography of Disaster: 'Vulnerability' and 'Local Knowledge' in Western Discourse." In *Mapping Vulnerability. Disasters, Development and People*, eds. Greg Bankoff, Georg Farks and Dorothea Hilhorst, 25–36. London: Earthscan, 2004.

---

8 The concept of the counter-public was proposed by Nancy Fraser (1991) in her reformulation of Habermas's notion of the public sphere.

Barry, Andrew. *Political Machines. Governing a Technological Society*. London: Athlone Press, 2001.

Beck, Ulrich. *The Risk Society. Towards a New Modernity*. London: Sage, 1992.

Beck, Ulrich. *World Risk Society*. Cambridge: Polity Press, 1999.

Boltanski, Luc. *La souffrance à distance*. Paris: Metailié, 1999.

Boltanski, Luc. "The Legitimacy of Humanitarian Actions and their Media Representation: The Case of France." *Ethical Perspectives* 7, no. 1 (2000): 3–15.

Callon, Michel, and Vololona Rabeharisoa. "Gino's lesson on humanity: genetics, mutual entanglements and the sociologist's role." *Economy and Society* 33, no. 1 (2004): 1–27.

Cefaï, Daniel, and Dominique Pasquier. "Introduction." *Les sens du public. Publics politiques, publics mediatiques*. Paris: PUF, 2003.

Chambers, Simone, and Will Kymlicka, eds. *Alternative Conceptions of Civil Society*. Princeton: Princeton University Press, 2002.

Clarke, Lee. "Panic: Myth or Reality?" *Contexts*, 2002 (Fall): 21–26.

Clarke, Lee. *Worst Cases. Terror and Catastrophe in the Popular Imagination*. Chicago: University of Chicago Press, 2005.

Cook, Deborah. "The Talking Cure in Habermas's Republic." *New Left Review* 12, no. 2 (2001): 135–151.

Davis, Mike. *Ecology of Fear. Los Angeles and the Imagination of Disaster*. New York: Metropolitan Books, 1998.

Dewey, John. *The Public and Its Problems*. Athens: Swallow Press, 1991.

Dupuy, Jean-Pierre. *Pour un catastrophisme éclairé. Quand l'impossible est certain*. Paris: Seuil, 2002.

Dupuy, Jean-Pierre. *Petite métaphysique des tsunamis*. Paris: Seuil, 2005.

Fraser, Nancy. "Rethinking the Public Sphere: A Contribution to the Critique of Actually Existing Democracy." In *Habermas and the Public Sphere,* ed. Craig Calhoun, 109–142. Cambridge, MA: MIT Press, 1991.

Furedi, Frank. *Culture of Fear. Risk-Taking and the Morality of Low Expectation*. New York: Continuum, 2002.

Furedi, Frank. *Politics of Fear. Beyond Left and Right*. New York: Continuum, 2005.

Giddens, Anthony. *The Consequences of Modernity*. Oxford: Polity Press, 1991.

Giroux, Henri. *Stormy Weather. Katrina and the Politics of Disposability*. Boulder, CO: Paradigm Publishers, 2006.

Habermas, Jürgen. *The Structural Transformation of the Public Sphere. An Inquiry into a Category of Bourgeois Society*. Cambridge: Polity Press, 1989.

Habermas, Jürgen. *Between Facts and Norms. Legitimizing Power?* Cambridge, MA: MIT Press, 1996.

Hood, Christopher, Henry Rothstein, and Robert Baldwin. *The Government of Risk: Understanding Risk Regulation Regimes*. Oxford: Oxford University Press, 2001.

Jasanoff, Sheila. "Technologies of Humility: Citizen Participation in Governing Science." *Minerva* 41, no. 3 (2003): 223–244.

Jasanoff, Sheila. *Designs on Nature. Science and Democracy in Europe and the United States*. Princeton: Princeton University Press, 2005.

Klinenberg, Eric. *Heat Wave: A Social Autopsy of Disaster in Chicago*. Chicago: University of Chicago Press, 2002.

Lakoff, Andrew. "Preparing for the Next Emergency." *Public Culture* 19, no. 2 (2006): 247–271.

Latour, Bruno. *Reassembling the Social. An Introduction to Actor-Network-Theory*. Oxford: Oxford University Press, 2005.

Lee, Spike. *When the Levees Broke. A Requiem in Four Acts*. DVD. New York: Home Box Office, 2006.

van Loon, Joost. *Risk and Technological Culture. Towards a Sociology of Virulence*. London: Routledge, 2002.

Luhmann, Niklas. *Risk: A Sociological Theory*. Berlin: Walter de Gruyter, and New York: Aldine de Gruyter, 1993.

Mignolo, Walter D. "On Subalterns and Other Agencies." *Postcolonial Studies* 8, no. 4 (2005a): 381–407.

Mignolo, Walter D. "Cambiando las Éticas y las Políticas del Conocimiento: La Lógica de la Colonialidad y la Postcolonialidad Imperial." *Tabula Rasa*, no. 3 (2005b): 47–72.

Negt, Oskar. *L'espace public oppositionnel*. Paris: Payot, 2007.

Neumann, Alexander. "Oskar Negt et le courant chaud de la théorie critique: espace public oppositionnel, subjectivité rebelle, travail vivant." In *L'espace public oppositionnel*, ed. Oskar Negt, 7–23. Paris: Payot, 2007.

Santos, Boaventura de Sousa. *Towards a New Common Sense. Law, Science and Politics in the Paradigmatic Transition*. New York: Routledge, 1995.

Santos, Boaventura de Sousa. "Between Prospero and Caliban: Colonialism, Postcolonialism, and Inter-Identity." *Luzo-Brazilian Review* 39, no. 2 (2002): 9–43.

Sloterdijk, Peter. *Le palais de cristal. À l'intérieur du capitalisme planétaire*. Paris: Maren Sell Éditeurs, 2006.

Somers, Margaret. "Narrating and Naturalizing Civil Society and Citizenship Theory: The Place of Political Culture and the Public Sphere." *Sociological Theory* 13, no. 3 (1995): 229–274.

Thévenot, Laurent. "Faire entendre une voix. Régimes d'engagement dans les mouvements sociaux." *Mouvements* 3, March-April (1999): 73–82.

Twigg, John, and Diane Steiner. "Mainstreaming Disaster Mitigation: Challenges to Organisational Learning in NGOs." *Development in Practice* 12, no. 3–4 (2000): 473–479.

Internet:

Beck, Ulrich (2001). "The Cosmopolitan State. Towards a Realistic Utopia." *Eurozine*. http://www.eurozine.com/articles/2001-12-05-beck-en.html (accessed 26 August 2014).

Philips, Peter (2004). "Tsunami Disaster Highlights Corporate Media Hypocrisy." *American Patriot Friends Network*, 30 December. http://www.apfn.net/messageboard/01-10-05/discussion.cgi.54.html (accessed 26 August 2014).

Walia, Harsha (2005). "The Tsunami and the Discourse of Compassion." *Seven Oaks*, Vol. 2. 1: January 4. http://www.sevenoaksmag.com/features/45_feat1.html (accessed 27 August 2014).

Carmen Diego Gonçalves

The Importance of Social Capital Facing the Unexpected (from Natural Hazards to Social Disasters): A Style of Thought[1]

1 The Idea of Social Capital in Disaster Definition

Disasters, as a concept, involve nominal and subjective definitions formulated by individuals and social entities. They are disruptive of social systems at small to more inclusive levels and are intertwined with broader dynamics of change. Capturing the antecedents and consequences of disasters is part and parcel of constructing descriptive and explanatory models of hazards and disasters. Classification schemes have frequently been based on defining the respective characteristics of disasters such as their length of forewarning, detection ability, speed of onset, and magnitude, scope and duration of impact.

Focusing our discussion requires sorting out the various ways in which a "disaster" is defined. Firstly, the conventional media definition centers on the destruction of human capital. Secondly, the focus falls on the destruction of physical capital. According to the Center for Research on the Epidemiology of Disasters (CRED), the University of Leuven (Belgium) during the period between 2000 and 2009: (a) about 4,000 disasters were recorded; (b) more than 780,000 people died; (c) more than 2 billion people were affected; and (d) the economic losses were estimated at USD 960 billion.

However, a third facet that we would like to highlight here involves social capital and, consequently, approaches disasters as a threat to existing social capital as both a predictor and a protective factor. This definition emphasizes how a disaster is a normatively defined situation in a community when extraordinary efforts are taken to protect and benefit social resources whose existence is perceived as threatened (Dynes 1998). The focus is on the community, as a social system, with concerns targeting response phase, which cannot be sepa-

---

1 This work was carried out under a post-doctoral research grant from FCT. The text was presented as a paper at the "Hazardous Future: Disaster, Representation and the Assessment of Risk" conference, Arrábida, 10–12 November 2011. The problems here discussed were also partially presented at Câmara Municipal da Amadora and Sociedade de Geografia de Lisboa (12 and 13 October, 2011, respectively), and to the Thematic Section of Knowledge, Science and Technology of the Portuguese Association of Sociology (18–19 November, 2011).

rated from the governance characteristics of what were once called developed societies and sustainable development.

The most difficult to study facet of disasters is the response phase primarily because there is frequently a shortage of time to cope with the often unexpected happenings. This makes disasters difficult for "planned" research. Some "response" research takes place months and years later and raises questions about the nature of recall and perhaps more importantly misses out on the emergent qualities of the response (Quarantelli 1997).

In *Organized Behavior in Disaster* (Dynes 1970), conceptualizations from fieldwork serve as the basis for their re-conceptualization in terms of social capital. This furthermore demonstrates how community social capital is the primary basis for resilience. The concept of social capital/interaction networks (Bourdieu 1983/86) was highlighted in the framework of disasters. Social capital refers to aspects of social structure, which are of value to social actors as resources susceptible to mobilization in pursuit of their interests. This is not located in the actors themselves as with human capital and instead located in the relationships and personnel networks between and among social actors. Social capital represents a resource embedded in the social structure that provides assets for individual and collective action. For social capital to hold value, there must be trust that the resources will be there to be drawn on when needed.

The concept of social capital has been deployed in the analysis of many different collective action problems, including family issues, schooling and education, work and organizations, democracy and governance, as well as development issues (Woolcock 1998). Coleman (1990) suggests that *physical capital* is created by making changes in materials so as to form tools that facilitate production, *human capital* arises out of changing persons to endow them with skills and capabilities, while *social capital* stems from the relationships among persons changing in ways that facilitate action. Physical capital is wholly tangible and embedded in observable forms. Human capital is less tangible as embedded in skills and knowledge whilst social capital proves even less tangible given its embeddedness in relations among persons. As such, the value of the concept resides in how resources are able to be combined in forms that account for the outcomes of social systems.

In turn, Coleman (1990) identified six forms of social capital. (1) *Obligations and expectations*: any social system relies heavily on reciprocal actions and rates obligations and expectations on the part of participants. Differences in social structures with respect to the extent of outstanding obligations arise for different reasons – trustworthiness that leads obligations to be repaid, the actual needs persons have for help; the existence of other sources of aid, the degree of affluence, cultural differences in the tendency to lend aid and seek aid. The extent of outstanding obligations within a system represents a measure of its interconnectedness as members remain obligated to one another. This connectedness also increases the resources available to each member. (2) *Informational potential*: information is important to provide a basis for action. Information results from recourse to social relationships maintained for other purposes. By interacting with informed members, individuals may increase their knowledge without having to obtain that information directly. (3) *Norms and effective sanctions*: norms support and provide rewards for specific behavior. Norms that encourage the subjugation of self-interest to the needs of the community are an especially powerful form of social capital facilitating certain actions while constraining others. (4) *Authority relations*: within groups organized to accomplish specific goals, a leader is often chosen and given the right to make decisions and speak for the group. In this transfer to one individual, the leader has access to an extensive network of social capital, which amplifies the social capital of individual members. In certain situations, the vesting of leadership gets attributed to a charismatic figure. (5) *Appropriable social organizations*: social organizations are usually created for a particular purpose and after that purpose has been resolved, the organization may redefine their goals. Thus, an organization developed for one purpose also serves other purposes – sometimes for short-term purposes and, at other times, for longer-term transformations. And, (6) *Intentional organizations*: developing social organization requires investment in designing the structure of obligations and expectations, responsibility, authority, norms, and sanctions. Creating such social organization not only advances the interests of those investing in it but also creates a public good benefiting others who do not invest directly.

Studies conducted from the sociological perspective of disasters return evidence that people associated in social networks during peri- and post-event are more resilient in recovery (Dynes 2005; Quarentelli 2000; Aguirre et al. 1995).

2 Natural Disasters and the Sociology of Risk and Uncertainty

So called natural disasters happen when physical factors affect populations and infrastructures, whose vulnerability index facilitates the destructive effects of those phenomena ("Natural hazards on their own do not result in disaster", Disaster Risk Reduction – United Nations Development Programme). It is important to analyze not only the physical and environmental factors triggering several natural phenomena but also the socio economic conditions of the prevailing contexts, the productive systems of goods and services and the perceptions and behaviors incising on the vulnerability index and consequently on increasing the destructive effects stemming from natural phenomena.

Natural disasters must be understood as the consequences of the impact of a natural event or hazard on a socio-economic system, with a given level of vulnerability, which may prevent the affected populations from coping adequately with the impact of more or less unexpected situations. Disasters can be catastrophic, depending on the magnitude of their massive destructive impact (Costa Maia 2007). The majority of victims from catastrophes occurring at a world level are from subdeveloped countries, where the lack of resources and high levels of vulnerability increase the scope of impact of several natural phenomena. Socio-economic factors, such as poverty, urban concentration and environmental degradation play a determining role in the consequences produced by natural phenomena.

Before, during and after they occur, disasters are physical and social catalysts of collective action (Tierney et al. 2001). Research on hazards and disasters requires an appeal to the scientific logic of discovery and explanation, regardless of substantive topic and whether the research is discipline, multidisciplinary, or interdisciplinary based.

The data available (Diego Gonçalves 2005) show that the conceptions of earthquake risk among scientists are related with styles of thought anchored on non predictable or indeterminate presuppositions. However, the physical consequences are predictable given the knowledge on potential sismogenetic sources that may affect an area and the physical characteristics of that same area. Hence, while it is not possible to prevent earthquakes from occurring, their effects can be significantly minimized through effective prevention measures and the reduction of vulnerability. However, the high margin of uncertainty or even indeterminacy related with human behavior also proves unavoidable. Uncertainty increases from nature to human behavior. Thus, the uncertainty dimensions concerning the individual and group behavior require consideration through analyzing the variables influencing vulnerability and interrelated with psychological and socio cultural factors within the context of risk exposition and primarily deriving from dimensions of knowledge and perception. Those factors are inculcated into individuals, civil society and its organization and span its normal life as well as exceptional emergency or crisis situations.

The consequences of the probability of hazard and uncertainty focus a generalized need for responsibility to enable development programs and models based on communication processes in communities ensuring prevention and even predictability as regards the damages arising out of natural phenomena (Diego Gonçalves 2006). Within this perspective, natural phenomena span the scope of natural disasters because they constitute an acceleration of the velocity and magnitude of the vulnerability of modern societies, the so called "risk societies". Accordingly, the Beck (1999) theory of risk societies proposes an image making the circumstances of modernity contingent, ambivalent and (involuntarily) susceptible to political rearrangements. Wisner and colleagues (2004), in the "Crunch" Pressure and Release (PAR) Model, argue that a disaster is a product of social, political and economic frameworks (facing the natural event – hazard) that are distinct from and should not be confused with the natural environment. Consequently, the social vulnerability conditions represent the root of the natural phenomena risk as a disaster problem and thus appealing to decision-making processes.

3 At Risk

States and citizens face dangers and risks that become more systematically intensified whenever the level of vulnerability and uncertainty associated with decision-making processes rises (Wynne 2002). As Luhmann (1993) points out, the difference between danger and risk is related to the fact that a danger is something to which people are exposed without having taken a decision, while risk is associated with the decision to take that risk. Hence, people may end up exposed to the consequences of decisions of someone else, such as technical and political decisions.

However, we consider that there is no unequivocal separation between danger and risk. In the risks that societies face (Beck 2000), experience is relevant for decisions involving the future even though contexts do change and new factors may arise affecting decisions about risk, whether individual, institutional or political. Thus, decisions must often be made under conditions of high uncertainty. More precisely, as pointed by Kasperson (2009), "deep uncertainty" where alternative approaches to risk analysis and for coping with this uncertainty must be found. Some hazards are more uncertain than others and human behavior also represents another factor of uncertainty – deep uncertainty proves a good concept for analyzing the vulnerability and resilience processes co-related with both nature and human behavior within the disasters problematic.[2]

---

2 Uncertainty is an inescapable ingredient of life. Even in everyday life situations – such as crossing a street – some level of uncertainty inevitably exists. Past experience is relevant for all decisions involving the future but contexts change and new elements affecting risk may unex-

It has become conventional to categorize disasters along the time dimension of preparedness, response, recovery and mitigation interrelated with a fundamental distinction between Hazard and Disaster Research around the risk dimension. The distinction between hazard, risk and disaster is important because this illustrates the diversity of perspectives on how we recognize and assess environmental threats (hazards), what we do about them (risks) and how we respond to them after they occur (disasters) (Cutter 2001). Therefore, this also reflects the different disciplinary orientations of researchers and practitioners.

Risk of a disaster relates with the probability of a disaster occurring. For management reasons, risk has been defined as the economic, social and environmental conditions resulting from hazardous events over a specific period of time. However, as the nature of hazards, risks and disasters became more complex and intertwined and the fields of hazard research and management more integrated, these distinctions became blurred as did the differentiation between the origins as "natural," "technological" or "environmental." The blurred distinctions highlighted by Cutter (2001) have contributed greatly to breaking down historical barriers between hazard and disaster research. The relationship between hazard and vulnerability leads to risk management – a tool to identify, analyze and quantify the potential damage of hazards and the actions required to reduce vulnerability.

Risk factors may be cumulative and carry additional and exponential risks when co-occurring. For example, poverty and low socioeconomic status are coupled with lower academic achievement and more emotional or behavioral problems. The evaluation of a risk should include vulnerability assessment and impact prediction taking into account thresholds that define the risk acceptable to any given society. Concomitantly, social dimensions of risk also depend on social, cultural and political values, such as equity, control, trust, visibility, transparency, accountability and public participation (Wynne 1996).

An earlier study (Diego Gonçalves 2009) concludes that, in the areas of trust, regulation, transparency, accountability, it is possible to identify assumptions shared between specialists and laypeople; there are common patterns of thought, anchored in shared assumptions, the *themata*, that reveal a style of thought – that we call the Paradigm of Accountability. Hence, shared patterns, prevention, preparation and emergency management practices cannot be out-

---

pectedly emerge. Usually, this residual uncertainty remains within reasonable boundaries and human beings make their way in the uncertain and changing world where existing knowledge and experience suffice as a guide to future expectations.

lined separately from the scientific capability of seismic risk prevision or the public's perception of risk.

Risk perception, in turn, should be understood from a multidisciplinary perspective related with several factors: information coming from society, the type of society, life styles and quality of life as well as global activities, the reproduction of social life, satisfaction of citizen necessities, the cultural ambience as well as factors such as: personality, personal and/or group experiences, disposition for risk taken, school and socio-economic levels, among others. Furthermore, behind these preferences, there are institutions and behind those there come cultures, social constructions and collective representations with their own logics (Douglas 1992; Grove-White et al. 2011; Renn 1992).

The "real" consequences of the risks lay people accept and/or are willing to accept are always blown up by social and cultural interpretations in turn dependent on the prevalent values and interests. People may feel predisposed to accept risks when they feel their objectives justify those risks. However, they can simultaneously reject any chance of incurring damages whenever feeling risks imposed upon them or feeling they run counter to their convictions and values (Slovic 1992).

Risk assessment requires a multidisciplinary evaluation taking into account not only the physical damage expectations, the number and type of casualties or economic losses (direct impact) but also the conditions related to social fragility and lack of resilience, which favor the second order effects (indirect impact) following a dangerous event (Carreño et al. 2007). Therefore, in a world of complex systems involving highly coupled human and natural systems and multifaceted social, economic and political institutions, high levels of uncertainty challenge existing assessment methods and established procedures for decision-making and risk management (Kasperson 2009) that necessarily requires investment in community resilience.

4 Risk, Resilience and Vulnerability

The phrase "risk and resilience" is quite common in this area of study. Indeed, if vulnerability can be defined as an internal risk factor, vulnerability and resilience, both collective and individual, represent key dimensions to the sociopsychological counterpart of the exposure to several stressors, describing the degree to which a social context and the individuals in it are susceptible to the short and medium-term effects of traumatic events and also convey how such effects may be overcome and positive reconstruction accomplished alongside better prevention (Diego Gonçalves and Possolo 2012).

Resilience is defined either as the capacity to achieve positive results in high risk situations or to maintain competencies whilst under threat or even the facing of unexpected or low probability occurrence situations such as natural disasters or other stressors; and making them favorable in order to recover from traumas. Resilience also serves to indicate a characteristic of resistance to future negative events. In this sense "resilience" corresponds to cumulative "protective factors" and is deployed in opposition to cumulative "risk factors". Researchers endeavor to uncover how some factors may contribute to positive outcomes, which are related to the protective factors explaining positive people's adaptation to adverse conditions. Positive adaptation, in turn, is considered a demonstration of manifested behavior of social competence or success at meeting any particular task at a specific life stage, such as the absence of psychiatric distress after exposure to those events. Concomitantly, a resilient attitude interrelates with adaptation capacities and recuperation faculties. In other words, one can say that resilience adds to the capacity of positive reconstruction in overcoming problems of re-mining, and cognitive flexibility. Resilient people are expected to adapt successfully even though they experience risk factors that run counter to good development.

However, being resilient does not mean being non-vulnerable. The degree of vulnerability is determined by a combination of factors including: (a) knowledge about hazards; (b) the conduct and behavior of populations and infrastructures; (c) public policy and management; (d) organizational skills in all fields of disaster management; (e) a certain degree of uncertainty not only in nature and scientific knowledge but also in the social system.

Interpreting the multidimensional concept of vulnerability, Andrew Maskrey (1997) states that the vulnerability of a community is expressed through many factors: lack of awareness or knowledge of the behavior of threats (cultural vulnerability); counterproductive legal, regulatory and institutional frameworks (institutional vulnerability); and disarticulations of social organization (social vulnerability).

Despite the considerable attention given to the research components of biophysical and environmental vulnerability (Mileti 1999), we still know little about the socially generated aspects of vulnerability, mainly due to the difficulty in quantifying them, which also explains why they usually do not appear in reports estimating post-disaster costs or losses. Cutter et al. (2008), applying the model of disaster of places, suggest that social vulnerability is a multidimensional concept that helps to identify those characteristics and experiences of communities (and individuals) that enable them to respond and recover from the consequences of disasters and hence not disconnected from the concept of resilience. And in that sense, social vulnerability is also not disconnected from the properties of resilience, such as: robustness: ability to withstand stress without degradation or loss of function; redundancy: substitutability of elements, systems and resources with respect to functional requirements; resourcefulness: ability to identify problems, to formulate priorities and apply resources to achieve goals; rapidity: ability to address priorities and accomplish goals in a timely manner so as to contain losses and prevent future disruption (Bruneau et al. 2003).

We may add that there is a widespread consensus in the social science community about some of the factors influencing social vulnerability, which includes: lack of access to resources (including information, knowledge and technology); limited access to power, political representation and legislation; lack of social capital, including social networks and connections, beliefs, customs; individuals with physical and/or mental disabilities, immigrants, lack of knowledge of the host country language, the homeless, temporary residents, tourists and type and density of infrastructure (Cutter et al. 2003; Tierney et al. 2001; Putnam 2000; Blaikie et al. 1994), which lead us to reflect about stratified vulnerability.

5 Stratified Vulnerability

While the functionalist approach that characterized classical disaster research mainly addressed the fact of disaster, not the sources of disaster vulnerability, other work has sought to better understand the societal processes that create vulnerability, how vulnerability is distributed unequally across societies, communities and social groups, how vulnerability changes over time and how and why these changes come about. The issues of structuring inequality and social stratification were long ignored in the first sociological studies conducted on the issue of disaster. It was the empirical data (unanticipated deductively) that began to accumulate evidence on inequalities in the behavior of populations during the peri-event and post-event recovery. Bates and colleagues (1963) had already discovered that individuals belonging to the working classes in the case of Hurricane Audrey suffered disproportionately high losses in relation to the upper middle class or upper class. Subsequently, several studies support the assumption of stratified vulnerability based on: racial and ethnic stratification, political power, gender, demonstrating how disasters exacerbate pre-existing inequalities (Bates and Peacock 2008; Oliver-Smith 1996).

Klinenberg (2002) deployed a social autopsy approach in order to illustrate how a disproportionate number of victims of heat waves mainly belonged to the elderly group of African-class Americans. In essence, this demonstrates how the social structure of a social context (Chicago, in this case) creates a distribution of victims stratified by race, class, gender and age. In the case of hurricanes Katrina or Rita, people with extensive social networks were able to turn to them to accommodate family and friends outside the impact zones, convert their capital and provide resources during the period of non-operability or destruction of their home areas in and around New Orleans (Barnshaw and Trainor 2007).

Trust, associations and norms of reciprocity between groups and individuals, including beliefs and customs, represent the capital that social networks visibly provide. Social and cultural capital, in the form of social networks, converts into tangible resources for the survivors of a hazardous event or disaster. Furthermore, these networks create close ties through identification with one's fellows also helping to reduce vulnerabilities.

6 Social Support in Disasters as Stressful and Traumatic Experiences

The broader social support literature suggests that perceptions of support networks may play a crucial role in determining the degree to which individuals are willing to seek out and utilize social resources. The inverse association between symptoms of post-traumatic stress disorder (PTSD) and social support, a means of tangible social capital, is one of the most consistent relationships observed in trauma research (Brewin et al. 2000; Ozer et al. 2003).

Recent meta-analysis (Ozer et al. 2003) indicates a robust association between PTSD and social support. This association has historically been interpreted within the framework of Cohen and Wills (1985) as stress-buffering model. More recently, trauma researchers have begun to explore an alternative model wherein the symptoms of PTSD contribute to the erosion of social support over time (King et al. 2006; Laffaye et al. 2008). The erosion model suggests that symptoms associated with PTSD negatively impact upon the quality and quantity of received support. What factors serve as intervening variables in the relationship between PTSD and social support? The isolation of relevant factors would bolster the net of relationships between PTSD and social support, which may in turn inform models regarding the causal nature of their association. In addition, clarification of the process through which PTSD and social support are interrelated may isolate specific cognitions, behaviors or environmental factors of potential use as targets in social intervention. This search for intervening factors in this relationship is a promising avenue of research with the intent to apply the concept of social support in exploring disaster response at the community level. Obviously, there will be differences in social capital among different societies and within the same society. These differences need exploring to render the concept relevant at a more general level. The discussion of social capital occurs within the context of a more inclusive social theory identifying the central elements of the historic transformation of social life, especially the decline of "primordial" institutions (the traditional family, for example) as the central element of social organization and the replacement of these institutions by purposely constructed organizations (as social interactions) (Coleman 1993), that act as protective factors.

Protective factors can be maintained before, during and after traumatic events. Social and also family support is a multi-sectoral factor given the roles played at different times. Social support becomes important to monitoring post-trauma reactions with people exposed to traumatic situations incurring a high risk of developing PTSD when social support is low (McNally et al. 2003). Another trauma exposure protection factor relates to the preparation for dealing with certain situations, constituting for the effect a variable of resilience (McNally et al. 2003). Thus, the training people undertake before, during and after these events proves crucial to information processing and the development of more effective coping strategies because it reduces uncertainty and increases perceived control and triggers and prepares appropriate automatic answers to manage or to cope with the situation (Shalev 1996).

In fact, if disasters provide evidence about the vulnerability of communities, cities and countries to danger and the severity of the impact on their economic performances and social welfare systems, social capital based community resilience, as social support for crises situations, demonstrates the ability to take deliberate, meaningful, collective action; proactive and reactive elements; fortifying against social concerns and nurturing the potential to grow from a crisis (Jewkes and Murcott 2009; Kulig 2000).

7 A Culture of Risk

The available information on the causes, consequences and risk variables derived from natural hazards and human vulnerability remains poor, insufficient and hardly fits the needs of citizen actions in the context of disaster.

To my knowledge, the concept of social capital – unavoidably interconnect-ed with social support and social networks – has not been greatly applied to disaster responses, which is a classic situation involving collective action for mutual benefit.

The intellectual roots of the concept are many and varied (Woolcock 1998; Portes 1998; Lin 2001). At this point, social capital theory is somewhat diverse in the emphasis different theorists apply. There is consensus that social capital consists of resources embedded in social networks and social structure and which can be mobilized by actors. However, there are differences in the con-tents of this concept. First, some focus entirely on social networks: others like Bourdieu (1983/1986) emphasize the production of "cultural capital" by group members. A second issue, interrelated with the first, deals with how social capi-tal should be measured. Now, measures of physical capital have been proposed but not without still considerable conceptual issues. A third issue is the choice of the dependant variable. For example, Putnam's (1993) treatment of social capital relates to the importance of civic involvement in creating good govern-ment and his idea of the historic decline in civic involvement.

Nevertheless, the right to a safer and better environment and quality of life is a growing expectation of populations. Understanding the interaction between hazards, exposure and vulnerability is crucial for effective disaster prevention. Building resilient communities should be a national disaster policy priority. Risk management must be framed in governance measures that constitute the capital increase in the social participation of people in the definition of ac-ceptable risk, setting styles and consequent political expressions of citizenship, building bridges between the expertise in hazards, mitigation measures, pre-paredness and recovery.

Within the development of a culture of risk, nurturing the perception in the population that the mitigation of the emergencies caused by disasters is not only a task for governmental organizations and private specialized entities but also depends on direct public participation, organizing their behaviors in the face of emergencies afflicting communities. Besides the physical conditions, scientific conceptions of risk, the scientific capacity of risk prevision and other technological aspects, we need to attribute more attention to citizens, within the social context – the subject and object of civil protection and disasters admin-istration. Insight into the way communities respond to disaster stems from uti-lizing the concept of social capital, such as social support and resilient ability.

It is hoped that interdisciplinary practices and specialized scientific studies contribute to changing attitudes in the area of political risk – reactive/emer-gency prevention policies – increasing the resilience of the social context, based on social capital as a social support in dealing with the natural events that can configure disaster. Community assessment contributes to community self-awareness, critical reflection, skill development and identifies avenues for resilience building, thus enhancing the community's ability to mitigate, prepare, respond and recover.

References

Aguirre, Benigno, Denis Wenger, Thomas Glass, Marceline Diaz-Murillo and Gabriela Vigo. "The social organizations of search and rescue: evidence from the Guadalajara gasoline explosion." *International Journal of Mass Emergencies and Disasters* 13, no. 1 (1995): 67–92.

Barnshaw, John, and Joseph Trainor. "Race, Class, and Capital amidst the Hurricane Katrina Diaspora." In *The Sociology of Katrina: Perspectives on a Modern Catastrophe*, eds. David L. Brunsma, David Overfelt and J. Steven Picou, 91–105. Lanham, MD: Rowman & Littlefield, 2007.

Bates, Fredrick, Charles Fogleman, Vernon Parentin, R.H. Pittman and G.S. Tracy. *The Social and Psychological Consequences of a Natural Disaster. A Longitudinal Study of Hurricane Audrey*. Washington, D.C.: National Academy of Sciences and Natural Research Council, 1963.

Bates, Frederick, and Walter Peacock. *Living Conditions, Disasters and Development: An Approach to Cross-Cultural*. Athens: University of Georgia Press, 2008.

Beck, Ulrich. *World Risk Society*. Cambridge: Polity Press, 1999.

Beck, Ulrich. "Risk Society Revisited: Theory, Politics and Research Programmes". In *The Risk Society and Beyond. Critical Issues for Social Theory*, eds. Barbara Adam, Ulrich Beck and Joost van Loon, 211–229. London: Sage, 2000.

Birkmann, Jörn. "Measuring Vulnerability to Promote Disaster-resilient Societies: Conceptual Frameworks and Definitions." In *Measuring Vulnerability to Natural Hazards. Towards Disaster Resilient Societies*, ed. Jörn Birkmann, 9–54. New Delhi, India: Teri Press, 2006.

Blaikie, Piers, Terry Cannon, Ian Davis and Ben Wisner. *At Risk: Natural Hazards, People's Vulnerability, and Disasters*. London: Routledge, 1994.

Bogardi, Janos and Jörn Birkmann. "Vulnerability Assessment: The First Step Towards Sustainable Risk Reduction". In *Disaster and Society – From Hazard Assessment to Risk Reduction*, eds. Dörthe Malzahn and Tina Plapp, 75–82. Berlin: Logos, 2004.

Bourdieu, Pierre. "The Forms of Capital." In *Handbook of Theory and Research for the Sociology of Education*, ed. John Richardson, 241–258. Westport, CT: Greenwood Press, 1983/86.

Brewin, Chris, Bernice Andrews and John Valentine. "Meta-analysis of risk factors for post-traumatic stress disorder in trauma-exposed adults". *Journal of Consulting and Clinical Psychology* 68, no. 5 (2000): 748–766.

Bruneau, Michel, Stephanie Chang, Ronald Eguchi, George Lee, Thomas O'Rourke, Andrei Reinhorn, Masanobu Shinozuka, Kathleen Tierney, William Wallace and Detlof Winterfeldt. "A Framework to Quantitatively Assess and Enhance the Seismic Resilience of Communities". *Earthquake Spectra* 19 (2003): 733–752.

Cardona, Omar. "Environmental Management and Disaster Prevention: Holistic risk assessment and management". In *Natural Disaster Management*, ed. Jon Ingleton, 151–153. London: Tudor Rose, 1999.

Carreño, Martha-Liliana, Omar Cardona and Alex Barbat. "Urban Seismic Risk Evaluation: A Holistic Approach". *Natural Hazards* 40, no. 1 (2007): 137–172.

Cohen, Sheldon, and Thomas Wills. "Stress, social support, and the buffering hypothesis." *Psychological Bulletin* 98, no. 2 (1985): 310–357.

Coleman, James. *Foundations of Social Theory*. Cambridge, MA: Belknap Press, 1990.

Coleman, James. "The Rational Reconstruction of Society." *American Sociological Review* 58 (1993): 1–15.

Costa Maia, Ângela. "Factores Preditores de PTSD e Critérios de Selecção em Profissionais de Actuação na Crise". In *Psiquiatria de Catástrofe*, coord. Luísa Sales, 263–276. Coimbra: Almedina, 2007.

Cutter, Susan. *American Hazardscapes: The Regionalization of Hazards and Disasters*. Washington, D.C.: Joseph Henry Press, 2001.

Cutter, Susan, Bryan Boruff and Shirley Lynn. "Social vulnerability to environmental hazards." *Social Science Quarterly* 84, no. 2 (2003): 241–261.

Cutter, Susan, Lindsey Barnes, Melissa Berry, Christopher Burton, Elijah Evans, Eric Tate and Jennifer Webb. "A place-based model for understanding community resilience to natural disasters." *Global Environmental Change* 18 (2008): 598–606.

Diego Gonçalves, Carmen. Estilos de Pensamento nas Concepções e Percepções de Risco. O Risco Sísmico em Portugal Continental. Da Previsão à Prevenção. *Doctoral Dissertation*. Lisboa: ISCTE, 2005.

Diego Gonçalves, Carmen. "Da indecidibilidade no falhamento e perigosidade sísmicas à incerteza social e percepção do risco", em colaboração com António Ribeiro. In *Encontro de Saberes. Três Gerações de Bolseiros da Gulbenkian*, eds. Ana Tostões, E. R. de Arantes e Oliveira, J. M. Pinto Paixão and Pedro Magalhães, 441–453. Lisboa: Fundação Calouste Gulbenkian, 2006.

Diego Gonçalves, Carmen. "Da previsão à prevenção: risco sísmico em Portugal Continental". *Segurança em Protecção Civil, Revista de Planeamento e Gestão de Emergência* 3, 13–21. Lisboa: Petrica Editores, 2009.

Diego Gonçalves, Carmen, and António Possolo. "Social and Physical Aspects of Seismic Risk". In *Risk Models and Applications. Collected Papers*, eds. Horst Kremers and Alberto Susini, 49–77. Berlin, CODATA-Germany, 2012.

Douglas, Mary. *Risk and Blame*. London: Routledge and Kegan Paul, 1992.

Dynes, Russell. *Organized Behavior Disaster*. New York: Lexington Books, 1970.

Dynes, Russell. "Coming to Terms with Community Disaster." In *What is a Disaster: Perspectives on the Question*, ed. Enrico Quarantelli, 109–126. London: Routledge, 1998.

Dynes, Russell. "Community Social Capital as the Primary Basis for Resilience", *Preliminary Paper* #344, 1–49. University of Delaware, Disaster Research Center, 2005.

Grove-White, Robin, Phil Macnagthen, Sue Mayer and Brian Wynne. *Uncertain World: Genetically Modified Organisms, Food and Public Attitudes in Britain*. Georgetown University: Bioethics Research Library, 2011.

Jewkes, Rachel, and Anne Murcott. "Community Representatives: Representing the 'Community'?" *Social Science & Medicine* 48, no. 7 (2009): 843–858.

Kasperson, Roger. "Coping with deep uncertainty: Challenges for environmental assessment and decision-making." In *Uncertainty and Risk*, eds. Gabriele Bammer and Michael Smithson, 337–348. London: Earthscan, 2009.

King, Daniel, Casey Taft, Linda King, Charity Hammond and Erika Stone. "Directionality of the association between social support and posttraumatic stress disorder: a longitudinal investigation." *Journal of Applied Social Psychology* 36, no. 12 (2006): 2980–2992.

Klinenberg, Eric. *Heat Wave: A Social Autopsy of Disaster in Chicago*. Chicago: University of Chicago Press, 2002.

Kulig, Judith. "Community Resiliency: The Potential for Community Health Nursing Theory Development." *Public Health Nursing* 17, no. 5 (2000): 374–385.

Laffaye, Charlene, Steven Cavella, Kent Drescher and Craig Rosen. "Relationships among PTSD symptoms, social support, and support source in veterans with chronic PTSD." *Journal of Traumatic Stress* 21, no. 4 (2008): 394–401.

Lin, Nan. *Social Capital: A Theory of Social Structure and Action*. Cambridge, U.K.: Cambridge University Press, 2001.

Luhmann, Niklas. *Risk: A Sociological Theory*. Berlin and New York: Walter de Gruyter, 1993.

Maskrey, Andrew. "Comunidad y desastres en América Latina: estrategias e intervención". In *Viviendo en Riesgo: Comunidades Vulnerables y Prevención de Desastres en América Latina*, ed. Allan Lavell, 14–39. Bogotá, Colombia: La RED, Tercer Mundo Editores, 1997.

McNally, Richard, Richard Bryant and Anke Ehlers. "Does early psychological intervention promote recovery from posttraumatic stress?" *Psychological Science in The Public Interest* 4 (2003): 45–79.

Mileti, Dennis. *Disasters by Design*. Washington, DC: Joseph Henry Press, 1999.

Oliver-Smith, Anthony. "Anthropological Research on Hazards and Disasters." *Annual Review of Anthropology* 25 (1996): 303–328.

Ozer, Emily, Suzanne Best, Tami Lipsey and Daniel Weiss. "Predictors of posttraumatic stress disorder and symptoms in adults: a meta-analyses." *Psychological Bulletin* 129, no. 1 (2003): 52–73.

Portes, Alejandro. "Social Capital: Its Origins and Applications in Modern Sociology." *Annual Review of Sociology* 24 (1998): 1–24.

Putnam, Robert. *Making Democracy Work*. Princeton: Princeton University Press, 1993.

Putnam, Robert. *Bowling Alone: Collapse and Revival of the American Community*. New York: Simon & Schuster, 2000.

Quarantelli, Enrico. "Panic Behavior: Some Empirical Observations". In *Human Response to Tall Buildings*, ed. Donald Conway, 336–350. Stroudsburg, PA: Dowden, Hutchison and Ross, 1997.

Quarentelli, Enrico. "Disaster Management, Emergency Management and Civil Protection. The Historical Development of Organized Efforts to Plan for and Respond to Disaster". *Preliminary Paper* 301. Newark, DE: Disaster Research Center, 2000.

Renn, Ortwin. "Concepts of Risk: A Classification." In *Social Theories of Risk*, eds. Sheldon Krimsky and Dominic Golding, 53–79. Westport, CT, and London: Praeger, 1992.

Shalev, Arieh. "Stress versus traumatic stress from acute homeostatic reactions to chronic psychopathology." In *Traumatic Stress: The Effects of Overwhelming Experience on Mind, Body and Society*, eds. Bessel A. van der Kolk, Alexander C. McFarlane and Lars Weisaeth, 77–101. New York: The Guilford Press, 1996.

Slovic, Paul. "Perception of Risk: Reflections on the Psychometric Paradigm". In *Social Theories of Risk*, eds. Sheldon Krimsky and Dominic Golding, 117–152. Westport, CT and London: Praeger, 1992.

Tierney, Kathleen, Michael Lindell and Ronald Perry. *Facing the Unexpected: Disaster Preparedness and Response in the United States*. Washington, D.C.: Joseph Henry Press, 2001.

Wisner, Ben, Pier Blaikie, Terry Cannon and Ian Davis. *At Risk: Natural Hazards, People's Vulnerability, and Disasters* (2nd ed.). London: Routledge, 2004.

Woolcock, Michael. "Social Capital and Economic Development: Toward a Theoretical Synthesis and Policy Framework". *Theory and Society* 27, no. 2 (1998): 151–208.

Wynne, Brian. "Misunderstood Misunderstandings: Social Identities and Public Uptake of Science." In *Misunderstanding Science? The Public Reconstruction of Science and Technology*, eds. Alan Irwin and Brian Wynne, 19–46. Cambridge: Cambridge University Press, 1996.

Wynne, Brian. "Risk and Environment as Legitimatory Discourses of Technology: Reflexivity Inside Out?" *Current Sociology* 50 (2002): 459–477.

Toshio Kawai

Big Stories and Small Stories after a Traumatic Natural Disaster from a Psychotherapeutic Point of View

1 Listening to Stories

The 2011 earthquake off the Pacific coast of Tōhoku, which occurred on 11 March 2011, was a tremendous disaster. It was the most powerful earthquake ever to be recorded in Japan with a magnitude 9.0. About 16,000 people died and 4,000 are still missing. More than 350,000 houses and buildings were half destroyed or totally destroyed. More than 22,000 ships were lost. The huge tsunami, following the earthquake, triggered at the epicenter about 70 km off the coast, caused unprecedented damage and victims. About 90% of victims were swept away by the tsunami and were drowned. Moreover, the shock and the tsunami destroyed several nuclear power plants in Fukushima, which led to the secondary disaster of radiation leakages from the plants. Given the nature of the series of tragic events that occurred, this disaster is appropriately called "the Great East Japan Earthquake".

After this tremendous disaster, not only rescue parties and materials but also psychological relief teams were sent to stricken areas. Many volunteer psychotherapists and psychiatrists travelled to the region in order to support refugees psychologically. Afterwards, the psychological relief work was mainly organized by the Association of Japanese Clinical Psychology, which has more than 20,000 members. However, there are various trials for psychological relief works from each organization and orientation. The Association of Jungian Analysts Japan (AJAJ) and the Japanese Association for Sandplay Therapy (JAST) organized a joint working committee for psychological relief work for earthquake victims. Sandplay therapy is a method invented by Dora Kalff who was strongly influenced by Analytical Psychology of C.G. Jung. It is very popular in Japan and JAST has more than 2,000 members. This is probably because Japanese people still partly live within a pre-modern worldview in which experience is not primarily situated within the human subject but in the things or in the nature; in this sense the Sandplay therapy fits nicely into the worldview of Ikebana and the Japanese garden tradition (Kawai 2010). I am the chair of this joint working committee and our activity is reported partly on the following website (http://iaap.org/). Our project has several focuses. One important concept is care of caretakers such as psychotherapists, nursing teachers and firemen. The reason for this type of indirect intervention is that our team comes from a distance and cannot be on site permanently. Such caretakers are supposed to be able to endure psychological difficulties for a certain period of time and await interventions. Our second point is to send a school counselor to the stricken areas since Sandplay is suitable for children. This project has led to the care of people based at schools because not only children but also their teachers and parents come to consult the school counselor. Moreover, we would like to continue this activity into the long term.

Those caretakers have to hold in the difficult experiences and stories without any outlet for a period of time. That could become an unbearable burden for them. Thus, it is important that those stories are shared in a protected circle. In this sense, the care of soul means the care of stories. Stories should be listened to, respected and shared. One important aspect of such an intervention is to avoid having caretakers keep their traumatic stories to themselves and to instead be able to share them within a safe circle of professionals. Our project does not try to teach people or even impose new methods on them to cope with psychological difficulties but tries to make protagonists out of people who suffer and to learn from them as well. Indeed, many caretakers are overwhelmed by what the various new methodologies and approaches have given them.

In fact, we did not go to the Tohoku region immediately after the earthquake but waited as there was a need to reflect on what was happening. The first set of email and fax inquiries about the damages and situations among members of both associations in the Tohoku region met with hardly any replies. However, the second set of inquiries that were sent out was met with many responses indicating how our members experienced the disaster and how they were struggling in their professional activities. So, we decided to visit the site at the end of April for the first time, just after the partial recovery of Sendai airport.

2 Life and Death

On arriving at Sendai airport, we were shocked by the piles of destroyed cars swept away and crushed by the tsunami. By seeing the wiped-out trees and buildings, we could notice that the tsunami came right up to the airport building and destroyed everything around.

Upon arriving at Sendai main station by bus subsequently, I had another strange feeling because I expected a heavily destroyed city, like Kobe city after the Great Hanshin Earthquake of 1995. At least seen superficially from outside, almost no sign of the earthquake was noticeable. The contrast to the condition of the area around the airport was drastic. Later, we visited the area directly hit by the tsunami near the coast. There were ruins of buildings, destroyed cars and fallen trees. But, on the other side of the highway, almost no damage could be seen because the tsunami was stopped by the highway bank.

Because most of the damage was caused by the tsunami in the case of this Tohoku earthquake, there was a clear contrast in the damage wrought along the path of the tsunami versus areas outside the tsunami of 1995. At the least, the damage on one side versus the other side of a road proved completely different.

This is also why there were many stories about life and death. Some people could narrowly escape with their lives, while others unfortunately died. Some schools had no victims, while other schools had several or many victims. More than twenty thousand people were killed; most of them by the tsunami.

Concerning life and death drama, I would like to report only one story told by my colleague, Yasuhiro Tanaka, who is active in our psychological relief work team. His mother-in-law, who lived in Ishinomaki, a port town devastated by the tsunami, was missing for a week after the earthquake and so we were afraid the worst might have occurred. But luckily she was found alive in a high school. Two days before the huge earthquake, there had been a relatively big earthquake of magnitude 7.2 in the same area. This was regarded as a major foreshock of the main earthquake. Her neighbor who was Korean and had not experienced a big earthquake before was terribly upset and visited her to ask what the matter was. His mother-in-law explained that this was the earthquake. Her neighbor thanked her for the explanation and promised her to escape together in his car when a big one should come since she did not have a car. When the big earthquake really occurred in the afternoon of 11 March, her neighbor came up to her after the first long shake and suggested she escape with him by car. They went by car to the nearest elementary school to take refuge but were refused entry because it was already full of people. This school, which was regarded as safe and a place for refuge was hit by the tsunami soon afterwards and many people died. Hence, a seemingly unfortunate rejection turned out to be fortunate for them. They went on by car but the road was congested because many people were trying to get up the hill by car. The tsunami was already coming from behind. The Korean neighbor did not know the roads but his mother-in-law did. She suggested to her neighbor to make a right turn on the next street. Because of this decision, they were just able to escape the tsunami and arrive at a high school that was serving as a refugee camp before staying there for a week. Coming back home, she found her house totally destroyed, full of water with the corpse of an unknown person floating inside.

This story shows how my colleague's mother-in-law was saved by several coincidences. Without these coincidences, she might already be dead. If there had been no preliminary earthquake, if her neighbor had not come up to her, if they had been accepted by the first school, if they had not turned right on the road ... This story is already dreadful enough. But there are many stories how a person lost his or her partner, children, parents, friends.

3 Coincidences

As I experienced the Kobe earthquake personally in 1995, I was aware of the role of coincidence in lives saved though it might not have been as dramatic in the case of the Kobe earthquake as events on 11 March 2011.

It was early in the morning of winter 17 January 1995. I woke up to pick up our one year old baby in the middle of the night from his bedroom because he was crying and took him to my bedroom. A couple of hours later, I suddenly felt as if I were in a sinking ship. There was neither right nor left, up nor down. Luckily, in the bedroom, there was only one very high and heavy piece of furniture, which left no space between it and the ceiling, and that prevented it from falling over, so we were spared any injuries. The building remained somehow intact but the inside of our flat was totally destroyed. Seeing that our baby's room had been completely destroyed, I wondered how fortunate we were that he had cried and been taken out of his bedroom in the middle of the night. After a while, a fire broke out next to our building.

In contrast to my story, a small child of one of my colleagues at Konan university was unfortunately killed by a falling heavy box. Hence, a week after the earthquake I organized a small dinner with all of my seminar students to engage in exchanges. They told us how they experienced the earthquake and how they were saved. It was impressive how some people were saved by pure luck. For example, one student had gone out to a 24-hour convenience store just before the earthquake. If she had remained in her room, she would probably have been killed, or at least heavily injured.

4 Experience Sharing: Psychological Time

Stories about life and death are not told soon after the disaster. Very often they need a certain lapse of time before people are psychologically ready to tell their personal stories of their experiences. But this timing is very often not respected by the mass media in its eagerness to find and report dramatic stories immediately following any disaster. And in these days of communication by blog and twitter, people tend to ignore their psychological timing and disclose their stories too early. Because we regularly visited the area hit by tsunami, we noticed a general flow of psychological time. When we visited in July, four months after the disaster, we had many reports from school counselors of children beginning to talk about their nightmares. I am not of the opinion that psychotherapists should focus on and technically "work out" these nightmares in such cases. In the case of therapy with schizophrenic patients, it is reported that they begin to have dreams related to their delusions and hallucinations when the condition is a little bit stable. Japanese psychiatrist Hisao Nakai (1974), who is famous for his therapy of schizophrenics, interpreted this change as absorption of delusions into dream. Delusions can now be objectively experienced and placed in the dream. The analogy of this process experiencing and telling nightmares should not be interpreted as the reviving of traumatic events. It rather means that they cease to have power and are not so threatening any more. Therefore, it is important to simply listen to the story and to relieve the person from the story.

In a school we visited regularly, one teacher told us in July that he had recently dreamt about the earthquake and tsunami and wondered why. And the principal of the school we had met several times before told us for the first time his experience of the tsunami in detail. Maybe his experience could be digested in ways that he could talk about it only after it was not so overwhelming for him. His story was as follows.

After the earthquake, he first led all the teachers and pupils into the sports hall of the school. However, as he began to feel that the hall might not be resistant enough against a possible tsunami, he wanted to evacuate them from the hall to the rooftop of the main school building. He went there with the vice principal to check it. Then, on their very way to the school building, the tsunami came all of a sudden, the vice principal behind was swept away, the principal ahead could barely escape and ran up the stairs to the rooftop of the school building. From the rooftop, he observed a scene like hell where many broken houses, cars and people were drifting away (I have to mention that in those videos on TV and YouTube, we saw corpses and floating people that were technically deleted beforehand). He was desperate because he was afraid that the hall would be destroyed and the children killed. However, luckily, the vice principal could hold on to something and was saved afterwards. Furthermore, the hall somehow remained intact against the tsunami so that no child was injured or killed.

As I said, I think such critical stories are told when they are no longer too overwhelming for the individual. Our experience with the children at this school supports this hypothesis. In one of the first grade classes, by chance, children freely drew a picture in February and so just before the earthquake. Then, the teacher let them again draw a free picture in April. Most pictures portrayed negative effects of the earthquake and tsunami. It is especially impressive that the structure of the pictures was very often destroyed, which seemed to mean the psychic structure of children was fundamentally shaken by the earthquake and especially by the tsunami. According to picture drawing test and therapy, such disturbances in the structure of drawings is equivalent to that symptomatic of a psychotic crisis.

But in the pictures drawn again in June a clear recovery was noticeable. The structure of pictures no longer showed any disturbance. Thus, we may conclude that the experience of earthquake and tsunami indeed leaves a tremendous influence on the psychic structure of people after a short period. But most children, most people seem able to recover from it. Correspondingly, frequent reports of so-called traumatic experiences and nightmares in July, which is four months after the disaster, do not reflect the onset of traumatic complication. It means rather that the first psychological crisis is over. Although those nightmares and fearful stories are told after such stories become less overwhelming, it is still very important that those stories are shared by a receptive person and in a protective atmosphere.

5 Psychotherapy and Small Stories

The readers may imagine how often traumatic experiences of earthquake and tsunami would be told in psychotherapy. However, as we have experienced in many interventions after other natural and social disasters like typhoons, criminal attacks and suicides, most people begin to talk about something other than the traumatic events after a while. As I suggested before, it takes about four months after the earthquake disaster to overcome the initial shock. Very often adults and children start to speak about other problems like difficulties in the family relationship. So, the improvement in psychotherapy is brought on by dealing with relationships in the family.

Our working committee sent a school counselor to an elementary school hit by the tsunami. But according to the recent reports from the counselor, the earthquake and the tsunami as such are not the theme of counseling anymore. There are various problems which have very little to do directly with the earthquake though they could be caused indirectly by the earthquake disaster. We also have some reports from nursing teachers that they are anxious about some children and suspicious about the negative effects of the earthquake. Their psychological and behavioral problems are worthy of handling even while having very little to do with the earthquake directly.

Hence, the success of psychotherapy consists in the shift of stories, from the big story of the traumatic event to ones about small, personal problems. The big story is repressed and replaced, to use Freudian term, by small stories. This is also valid for the psychotherapy of so-called psychosomatic symptoms. I work as psychotherapist in a hospital specializing in thyroid diseases. If a medical doctor there diagnoses that psychological factors play a big role or a patient will only be cured exclusively by medical treatments with difficulty, he or she sends a patient to psychotherapy. Also in this case, if the patient talks only about physical problems and complaints, there is no chance for improvement. But psychotherapy and even physical recovery can be successful if the patient starts to talk about, for example, her mother-in-law, discontent with her husband and so forth. I would like to even suggest that some successful patients can make use of disaster for his or her psychological developments. A very impressive paper on psychological work with a boy after the earthquake in Northridge reported by Taki-Reece (2004) can be understood not as trauma work, but as psychological development drawing upon the crisis of the earthquake. The separation from his parents and establishment of his psychic center was the theme of the story expressed by his Sandplay. Psychotherapy cannot cure or create the big story but rather has to do with the small ones.

But the importance of small stories should be stressed. The Internet is overflowing with small, and I am inclined to say even unnecessary, stories in the form of personal blogs and twitter postings. I'm getting up. I am now eating. Sit on a bus heading to … But just after the earthquake, all such stories disappeared from the Internet. People were compelled, or at least felt compelled, to make a headline message "I feel deepest sympathy for those hit by the earthquake" and refrain from reporting daily, trivial events. One big story could dominate and suppress the small stories, which can also be unhealthy. Control of mass media by the government was a critical theme after the earthquake, which was also the theme of other papers in this volume, but we have to note that we also have a hidden agenda of self-control.

6 The Birth of New Big Stories

Psychotherapy has little to do with the big story but tries to help people find out small stories to live and cope with. When a big story is brought up in therapy, it is very often a sign that this story is about to end or one still sticks to the old story. Very seldom does a psychotherapist encounter a creative big story of the type which seems the preserve of artists and novelists. In this case, the big story tends to be symbolized and metaphorized in a work of art or novel instead of being replaced by small stories. Although psychotherapy has mostly to do with small, personal stories in manifest works, it is important to be aware of the role that the big story plays. Furthermore, not so much in an individual piece of work but in psychotherapy, we have to be concerned with the care of the collective soul and in the role of the big story in the collective soul.

I would like to mention only two of those stories based on previous earthquakes. One is from stories to live and cope with. One big story emerges out of a selection of tales from Tono region in Tohoku, written by Kunio Yanagita (1875–1962). Yanagita is the founder of Japanese folklore studies and this work has become a Japanese folklore and literature classic. This is a very interesting and peculiar work, which led to many studies and still deserves future studies from new vantage points. I would like to mention only one story which is related to a big earthquake. The coast of Sanriku was hit not only this time but also on several times by terrible tsunamis. Historically, the one in 869 seems the biggest with a tsunami 30 m above the sea. In 1896, there was the famous Sanrikuoki earthquake, which caused a tsunami and killed more than twenty thousand people. The story I quote, Nr. 99, is based on this tragedy. I would like to thank Prof. Sukeyuki Miura whose paper in a seminar reminded me of this story and its relating materials.

This story may be an answer to the life and death question.

99. Kiyoshi Kitagawa, an assistant headman in Tsuchi-buchi village, lived in Hiishi. His family had been roving priests for generations. His grandfather, named Seifuku-in, was a scholar who had written many books and done a lot for the village. Kiyoshins younger brother Fukuji married into a family in Tanohama on the coast. Fukuji lost his wife and one of his children in the tidal wave (tsunami) that struck the area last year. For about a year, he was with the two children who survived in a shelter set up on the site of the original house.

On a moonlit night in early summer, he got up to go to the toilet. It was off at some distance where the waves broke on the path by the beach. On this night, the fog hovered low, and he saw two people, a man and a woman, approaching him through the fog. The woman was definitely his wife who had died. Without thinking, he trailed after them to a cavern on the promontory in the direction of Funakoshi village. When he called out his wife's name, she looked back and smiled. The man he saw was from the same village, and he too had died in the tidal wave (tsunami) disaster. It had been rumored that this man and Fukuji's wife had been deeply in love before Fukuji had been picked to marry her.

She said, "I am now married to this man." Fukuji replied, "But don't you love your children?" The color of her face changed slightly and she cried. Fukuji didn't realize that he was talking with the dead. While he was looking down at his feet feeling sad and miserable, the man and the woman moved on quickly and disappeared around the mountain on the way to Oura. He tried to run after them and then suddenly realized they were the dead. He stood on the road thinking until daybreak and went home in the morning. It is said that he was sick for a long time after this.

We mentioned that the story about life and death was important in the case of a tsunami. How can the dead and living who were separated then meet again? Or how can the living say farewell to the dead? In Japan, there is a ritual to welcome dead ancestors during the Obon time in the summer. It is around the middle of August. In the case of the city of Kyoto, the dead come back between 7 and 10 August, and they stay around and leave this world on 16. The touristic famous sending fire of five mountains in Kyoto is a ceremony to send the dead soul back to the other world. In other places, the dead are sent back to the other world by floating dedicatory lanterns on the river. Many dedicatory lanterns on the Kitakami river near Ishinomki this year (2011) had a special meaning because they were not only for symbolic ancestors but for the real recent tsunami dead. This was no summer touristic festival but a real ritual.

Now story No. 99 says it takes place in the summer. In the text of *Legends of Tono* it is happening in the year following the Sanriku earthquake on the same day as the earthquake but in other texts based on the same colloquial tradition, they say this happens in the summer of the same year of the earthquake. Thus, this must be the first Obon for the dead to return home, which is very important. I have heard many dreams of patients who did not believe the return of the dead in Obon but was surprised to dream of the dead who returned to this world exactly during this period. So, Fukuji who had lost his wife met her again. This story is a verification of the old story and belief.

But this story is not idyllic and sentimental. It does not end up with a happy return of the dead. Fukuji lost her wife first physically in the tsunami and now loses her again psychologically because she seems to be living with her former lover. Meeting again leads to another separation. But this second loss is necessary to establishing a new relationship between life and death and to recovering from disaster. In the Greek mythology Orpheus goes to the Underworld to look for his dead wife Eurydice. He is allowed to take her back to this world with the promise of not seeing her on his way. But he looks back at her and loses her eternally. The Orpheus action should not be understood as a failure but as an accomplishment of love and sending Eurydice to her proper place since her name suggests that she is queen of the Underworld (Kawai 1989). In an analogy to the myth of Orpheus, our story from the *Legends of Tono* suggests that the dead wife is now psychologically lost and belongs to her own world. This only becomes possible by the dialectic encounter of life and death. Because this story is based on the Japanese tradition of Obon ritual and is similar to the Orpheus myth, this story can be regarded as a collective answer to the tragedy of tsunami. But it is at the same time Fukuji's own personal story. In this sense, this is an encounter of a big story with a small personal story.

I would like to introduce another story, *After the Quake*, written by Haruki Murakami. Haruki Murakami is probably the most famous contemporary novelist in Japan and is probably very close to the Novel prize. His recent novel, *1Q84*, was published in English in October 2011. In my understanding, his novel shows the state of the soul in Japan, which has skipped over the task of establishing modern consciousness and stands between the pre-modern world and postmodern consciousness (Kawai 2004). I would like to focus on the story "Superfrog saves Tokyo" which has a direct reference to the earthquake.

Katagiri, a banker, was surprised by a huge frog who was waiting for him in his house. The frog explained that there would be an earthquake in Tokyo at 8:30 am on 18 February caused by a huge worm, which was within three days of the frog's prophecy. The frog wanted to prevent the earthquake but needed help. The frog subsequently proved the accuracy of his prophecies to Katagiri by predicting a returned pending loan. The frog and Katagiri decided to go down to the epicenter located exactly beneath the bank where Katagiri worked. But, just the day before the earthquake, when they needed to go under the bank to fight against the worm, Katagiri was shot by someone and brought to a hospital. Katagiri woke up in the bed of the hospital and noticed that it was already 18 February and the earthquake had not happened. A nurse explained to him that he was found lying in the street and, contrary to his memory, without a gun wound. The frog came up to the hospital and explained how he, the frog, fought against the worm. Katagiri apparently was helping the frog in his dream. The frog started to fade as he was telling the story about the fight with the worm. Boils burst out on his body, and "wriggling, maggotlike worms of all shapes and sizes came crawling out". Hundreds of worms came and crawled up Katagiri's leg. He screamed and a nurse came up. He told her "He (the frog) saved Tokyo from being destroyed by an earthquake all by himself." The end of this story suggests that Katagiri was also about to die. "Then he closed his eyes and sank into a restful, dreamless sleep."

In contrast to *Legends of Tono* that depicts what happened after the earthquake, this story shows how the earthquake was prevented. It is interesting that Katagiri has "no wife, no kids, both parents dead, brother and sister he had put through college married off. So what if they killed him?" The lack of relationship makes him available to encounter the frog. In this sense, he does not live in the old view and story of meeting dead ancestors. While in the *Legends of Tono* the motivation is a personal one of meeting the dead wife, the matter here is the collective task of saving Tokyo from the earthquake and thus has a totally different dimension.

The frog fought against the worm while Katagiri wanted to help the frog and did help him in the dream. We can draw two conclusions. The first has to do with connecting with the source of power. The epicenter was located in the underground of the bank in which Katagiri worked. Is Katagiri a special person, a hero or a godlike prophet? No, this may suggest that every postmodern person who is not in a relationship has his epicenter in his own depth. It is important to come to contact with this depth in the form of fighting. I don't think the point is to overcome the danger and disaster. As the frog was transformed and died after the fight, he was sacrificed by coming into contact with the power of the Worm and was even embodied and transformed to worms. In the case of disaster, there are very often both aspects of fear and getting power from it. For example, on the coasts hit by the tsunami this time, there are hundreds of Shinto shrines. This is because people were both afraid of and thankful for the power of water, the power of the ocean which brings fish and other products. People were afraid of nature but, simultaneously, grateful for its richness.

These two attitudes, which are somehow equivalent to the feelings of *tremendum* and *fascinosum* which Rudolf Otto described as feelings toward the "holy", were historically noticed in other rituals. After many natural disasters and epidemics in Heian time in Japan, a festival began to drive away bad spirits. This was Aoi Matsuri in Kyoto. But according to historians, it was important for people at that time to get the power of bad spirits. So, there are both moments of fear and fascination.

If we analyze the attitude toward the nuclear power plant in Japan from this point of view, I have to say that it is totally out of this worldview. Nuclear power does not belong to environmental order and comes from outside of nature. No wonder Japanese have not built any shrines to give thanks to nuclear power plants and want to discard them now without any gratitude for such powers. There has not been a grounded story around the nuclear disaster but people believe that many of the stories around the nuclear power plant were false, manipulative and only served the interest of the Tokyo Electric Company, the government scientists and the mass media who colluded and spent billions of dollars on developing their version of the story surrounding the safeness and cleanliness of nuclear power plants.

I would like to come to the second conclusion. After the earthquake of 11 March many writers, including human scientists, highlighted the Japanese attitude of "Mujyo", a worldview of the calm and passive acceptance of fate, thinking that nothing remains the same, there is nothing in the end. But in Haruki Murakami's novel, no passive acceptance of the fate is recognized. Rather, the evil and the responsibility to fight the evil are recognized. This is because this work appeared as result of both natural disaster, earthquake and manmade poison-gas terrorism. We have to be active, have to fight against the evil powers. Because the earthquake of 11 March resulted in a combination of natural and manmade disaster, Haruki Murakami's novel somehow anticipated a new attitude. I would not say there is already a solution or even a hint of such. But we may at least say that the highly praised attitude of "Mujyo" among the Japanese after the earthquake 11 March is not valid anymore. We are in search of a new story.

7 Necessity of a New Story: Conclusion

Though I emphasized that psychotherapy consists of the emancipation from the big story and the creation of small stories, I may have overly devoted myself to the big story. But while I have worked with people in Tohoku, I have keenly felt the loss of the old story in the background. This loss overshadows the personal, psychological problems. So I feel the need for a new story.

But at the same time, it is absolutely necessary to shed light on how Tokyo Electric Company, the government, the scientists and the mass media colluded to create their version of the story and manipulate people. We have to learn from the negative consequences of this story.

There are several good ideas for a new story. For example, the work of Japanese anthropologist Shinichi Nakazawa, *Big Change of Japan* (2011) is worthy of mentioning. He analyzes the unmediated, immediate character of nuclear power. This does not belong to the environmental order and hence the Japanese worldview. But, instead of asserting a return to the old worldview and technology, he proposes direct but mediated use of sun energy, which is a dialectical negation of nuclear power.

However, it is still premature to speak about a new story concretely. It is now important to accept and carry the loss as loss. Because of the earthquake of 11 March, many lost loved ones, precious property and a place to live. As a psychologist and a psychotherapist, I would like to respect the loss of the old story and worldview so that the emptiness may become a place for the birth of a new story.

References

Kawai, Toshio. "Die Initiation ins Dichterische bei Heidegger und Jung: Der Ort der Psychotherapie." *Daseinsanalyse* 6, no. 3 (1989): 194–209.

Kawai, Toshio. "Postmodern consciousness in the novels of Haruki Murakami." In *The Cultural Complex,* ed. Thomas Singer, 90–101. London: Routledge, 2004.

Kawai, Toshio. "Jungian psychology in Japan: Between mythological world and contemporary consciousness." In *Cultures and Identities in Transition*, eds. Murray Stein and Raya A. Jones, 199–208. London and New York: Routledge, 2010.

Murakami, Haruki. *After the Quake.* New York: Alfred A. Knopf Publisher, 2002.

Nakai, Hisao. "Seishinbunretsubyou jyoutai karano kankaikatei" (Recovering process from schizophrenic conditions). In *Bunretsubyou no seishinbyouri 2*, ed. Tadao Miyamoto, 157–217. Tokyo: Tokyo University Press, 1974.

Nakazawa, Shinichi. *Nihon no Daitenkan* (Big change of Japan). Tokyo: Shueisha, 2011.

Otto, Rudolf. *Das Heilige: Über das Irrationale in der Idee des Göttlichen und sein Verhältnis zum Rationalen.* Breslau: Trewendt und Granier, 1920.

Taki-Reece, Sachiko. "Sandplay after a catastrophic encounter: From traumatic experience to emergence of a new self." *Archives of Sandplay Therapy* 17, no. 2 (2004): 65–75.

Yanagita, Kunio. *The Legends of Tono.* Trans. Ronald R. Morse. Lanham: Lexington Books, 2008 (1910).

Shoko Suzuki
In Search of the Lost *oikos*: Japan after the Earthquake of 11 March 2011

1 Japan after 11 March 2011

At 2:46 pm on 11 March 2011, north eastern Japan was hit by a magnitude 9.0 earthquake followed by a 7-meter tsunami. The death toll remains unknown. It seems certain to exceed 24,000 as whole sections of some communities were washed out to sea. Of that number, almost nine thousand remain unaccounted for. They were carried off by the tremendous wave that swept over the dikes. Most of those bodies have probably sunk to the floor of the cold sea. Search and rescue groups remain grimly at work finding bodies along the shore and beneath the rubble and debris.

The catastrophe that Japan suffered on that fateful day was not limited to earthquake and tsunami but also included explosions at nuclear power plants. In fact, this was a triple disaster: the earthquake itself, the devastating tsunami that followed, and the ensuing crisis at the Fukushima Daiichi Nuclear Power Station. The unprecedented natural disaster triggered violent explosions at the nuclear power plant. Radiation spread widely, a problem which has not been solved long after the accident. The human losses are already enormous and now the slow erosion of humanity threatens: there are sad reports of hundreds of elderly left to die in hospitals and care homes in the stricken areas.

Reactions to the disaster have revealed many aspects of Japanese society otherwise usually hidden in the course of ordinary daily life. These aspects now raise the following sensitive questions: Can scientists prevent the linkage between a natural disaster and a manmade one? Is there a sure way to distinguish between a natural disaster and a manmade one? How are the Japanese converting their experiences into knowledge and reflective thinking regarding their own society? What lessons are being learned and how? Can we share these lessons with the world? This essay focuses on these learning processes while describing the disaster from the viewpoint of a search for the lost *oikos*.

The magnitude of the disaster has freshly called into question the very foundations on which we have lived our lives and upon which we hope to continue to live in the future. This foundation is very literally the ground beneath our feet, the earth upon which we were born. At the same time, it is the *oikos* that can be our *topos* of peace and happiness.

Oikos is a Greek word that is the common root for the English words economy and ecology (Finley 1999: 14, 53/54, Schumpeter 1954: 54). It indicates a spatio-temporal realm linking the macrocosm of the world and universe with the microcosm of the human body. One might say that it is the human environment surrounding and embracing human beings and enabling them to relate to the wider world and the vastness of the universe. Rather than being a specific physical habitat in which people can live comfortably and pleasantly, this *oikos* is more of a "homeland of the heart," a richly textured spatio-temporal realm characterized more by quality than quantity. Moreover, the sense of being nurtured and protected by this *oikos* is fundamental to our shared sense or common sense of living in the here and now.

In the recent disaster people lost not only family members, homes, land and even the neighborhoods and towns in which they lived – they also lost their *oikos*. The psychic wound engendered by this loss is impossible to describe in words. How can this vanished *oikos* be restored? What role can and should culture play in providing the driving force for recovery from the disaster and in nurturing the capacity to envision and assess an uncertain future and respond appropriately to its risks?

2 Reinterpreting the Concept of *oikos* and Applying It to the Potential for Recovery

2.1 The Meaning of *oikos*

In ancient Greek, the word *oikos* originally meant a clan, a group of homes or a community. It was the fundamental unit of ancient Greek society and included the head of household, his wife and children, and any slaves living under the same roof. This conveys the focal point of our lives, the smallest unit of social organization. By extension, the word *oikos* was sometimes used to denote the possessions of the head of household, including the land cultivated by slaves as well as the slaves themselves (Vernant 1983, 130ff., 149, 161; Le Goff 1988, 20). Gradually, the scope of the word expanded from individual houses to the wider holdings of the head of household and, from there, to collections of homes, communities, and towns. Depending on the circumstances and context, the word could be used with a very broad meaning. Sometimes it was applied in situations where concrete boundaries could not be defined in which case it referred to a region or network of human bonds that were dedicated to upholding a common set of values. In this sense, it had the abstract meaning of a group with commonalities, or the place where such a group existed.

In her book *The Human Condition* (1958), Hannah Arendt makes a distinction between *polis*, which she defines as the public realm of the political community, and *oikos*, defined as the private realm of the household (Arendt 1958, 28–29, 33, 68–69). The former provides the basis for political, cultural and scientific activity, while the latter supports natural and life production. Her main focus is on the social realm, which she contends arose in the modern era as an intermediary domain between the private and public realms and encroaches upon those realms. Therefore, she does not fully articulate her perspective on such aspects as intimacy and affective bonds, which arise through activity in the private realm of *oikos*. Hints of the progression of her thought concerning a richer *oikos* are evident, however, in her analysis of Kant's conception of human nature found at the end of her late, unfinished work, *The Life of the Mind* (1978). Her analysis, which alludes to the human action of people as observed from the perspective of the earth and the universe, is guided in its development by her interpretation of St. Augustine, who believed that human beings are not an existence doomed to death but rather one of emerging birth (natals). People from several generations live together on a single planet they share and then die. It is with regret that I surmise that, for Arendt, the apprehension of this fact of human life connected *oikos* as a place to live with her concept of life activity (*vita activa*).

2.2 Being Alive in *oikos*

Being alive, and the quality of the act of living, affects whether or not a person experiences a simple place as being intimate. Simultaneously, human beings are organisms in a constant state of activity. If we consider the fact that, in the midst of this unstable activity, human beings derive stability at the center of the movements they make from moment to moment, the place that makes such human activity possible must itself be an organic entity.

Today, the two words economy and ecology refer to completely different academic disciplines even if deriving from the same root. We should not forget that economy (eco + nomos) originally meant the management of the home, while ecology (eco + logos) meant the logic of the home (Schumpeter 1954, 53; Finley 1999, 14). Indeed, we must consider it problematic that, under current circumstances, these two are understood to be completely separate fields in contemporary society. Thus, what is needed is a perspective that places logic and ordering on the same horizon.

From our perspective, the significance of ecological concepts lies in the realization that all of nature comprises a single organic system. This means that the various elements in an ecosystem are connected, creating a mesh of interrelatedness. Human beings exist in relation to eco and to other entities that share that eco, all of which coexist in a state of mutual influence. The relationship between human beings and eco (*oikos*) reflects the true meaning of the term as "something that belongs to me," or "something that has a connection with me."

In his book, *Being Alive: Essays on Movement, Knowledge and Description* (2011), Tim Ingold says that "being alive" is a process of becoming that has no goals; that is, the process itself creates the form. According to Ingold, the science of the past has viewed "being alive" as an output from piecemeal systems such as genetics/culture or nature/society. In this formulation, "being alive" appears to be nothing more than the act of filling these systems to a preset capacity. Ingold criticizes this conventional approach of science because it imposes a deterministic methodology on human beings who, by "being alive," are engaged in constant activity and transformation. So what, exactly, is "being alive"? Ingold contends that it means all individual living creatures inhabiting a specific environment with our bodies, living with our feet on the ground. In contrast to conventional science, which he criticizes because it seems to assume the existence of "life" as an abstract entity, he points out the importance of focusing on the manifestation of "life" in terms of individual, specific ecologies.

Ingold asserts the necessity of interpreting the human act of "being alive" in terms of "dwelling" (Ingold 2011, 9ff.) People walk the earth in specific environments. As they live, they discover and apply the meaning of things through individual actions performed amid constant change in light, wind and sound engendered by the weather. According to Ingold, this type of life is itself nothing other than the constant unfolding of the creative process. Incorporating the idea of ecological psychology espoused by James J. Gibson, he tries to define human activity, including cultural and social activity, as "dwelling." The concept here is that living as a kind of "dwelling" is not merely an abstract matter or mental representation but is actually led by the physical body, which acts and perceives. At first glance, Ingold's assertion appears to take us back to the pre-modern idea of a substantial or physical body and indeed his argument develops with constant reference to the thought of Deleuze and Guattari. Ingold attempts to combine Deleuze and Guattari's references to the body in the context of the handiwork of artisans (carpenters, craftsmen, metalworkers, masons, etc.), with his main thesis being that "being alive" is the same thing as "dwelling."

2.3 Being Alive as *poiesis*

Among readers who are even slightly familiar with the Kyoto School of Philosophy, Ingold's characterization of "being alive" as a goal-less creative process will no doubt bring to mind that school's founder Kitaro Nishida and his concept of *poiesis* (Nishitani 1991, 26). Nishida defines human life as an interaction between the environment/world and human beings. He contends that life is a creative process that changes and grows through the mutual influence of human beings and their environment. For him, the act of being alive for humans is truly an expressive and creative act. To illustrate this point, he discusses the creative process experienced by artisans as they make their works. He posits a creative process in which it is unclear whether the artisan is making the work himself or his body is being borrowed by a higher power that transcends the artisan (Elberfeld 1999).

During the creative process, a moment comes when the artisan finds it hard to tell if he himself is the agent making the work or if he is a medium through which some larger entity is making the work or if he is merely a channel through which the power of creation residing in the work itself is manifested. This is the moment when the muses smile and has been a part of the Western context since ancient Greece. Is the creator the subject or the object of creation? Is he neither or is he both? Subject and object are integrated to form a complete whole; it is as if they cannot be distinguished, or put another way, that they have not been separated or segmented in the first place. Nishida calls this state "pure experience" (Elberfeld 1999, 81–82; Heisig 2001, 42–43).

In this place delineated by Nishida where pure experience arises without a division between subject and object, one can discern a double creative process. On the one hand, an artisan creates a work as part of the creative process. On the other hand, the act of creating that work polishes his skills and his character as an artisan. In this sense, the artisan is actually creating his own life or, alternatively expressed, creating himself as a human being who is an artisan. The discovery of this two-fold creative process within the creative activity of an artisan is something that can often be found in Eastern thought. Nishida described it using the language of German Idealism in an attempt to depict a state in which subject and object are undifferentiated. He interprets the process of human life as a process through which people create and express themselves as works and defines "being alive" as *poiesis*.

If we apply Nishida's *poiesis* to the relationship between *oikos* and human beings, it becomes possible to construct the following four theses: 1) *oikos* is not a uniform, simple spatio-temporal realm as delineated by modern physics but rather is itself an organic system; 2) when human beings inhabit *oikos*, they also become organic systems that both create and are created by *oikos*; 3) the interaction between *oikos* and human beings is not quantitative in nature but qualitative; and 4) the relationship between human beings and *oikos* is such that each has a special meaning of "intimacy" for the other.

3 *Topophilia*, or Nostalgia for Home

Because *oikos* has a qualitative relationship with human beings as a place in time and space that has a special meaning, it also has an affective relationship with human beings. That affective bond is similar to nostalgia for home, or homesickness. In *Topophilia: A Study of Environmental Perception, Attitudes, and Values* (1974), the Chinese geographer Yi-Fu Tuan, who is based in the United States, considers the question of why people form fond attachments to particular environments or places by citing many examples from a historical and comparative cultural perspective. He coined the term "*topophilia*" to denote this fond attachment to place. *Topophilia* links places with emotions. Depending on the image created by a place or environment, people invest emotion in their relationship with it. Tuan delineates an emotional spectrum that begins with an aesthetic appreciation of a scenic spot or feelings of intimacy with a place, which leads to emotional or deliberate inclinations in space recognition and from there to nationalism. Concerning this series of emotions, Tuan elucidates the intimate feelings toward space that are deposited in historical materials, ethnography and literary symbols while tracing various sources such as perceptive and symbolic systems and symbolic acts. One common element he identified is a longing for natural beauty untouched by culture along with a nostalgic vision of home.

In the aftermath of the earthquake and tsunami, concerts were held by volunteers at evacuation sites, where the song "*Furusato* (My Old Country Home)" was frequently sung. This song, which is part of the elementary school music curriculum, is well-known to Japanese citizens of all ages. Let us examine the lyric.

> I chased rabbits in those mountains
> I caught minnows in those streams
> Even now I dream of it
> My unforgettable old country home

The scene painted here is that of a typical Japanese farming village. For the Japanese who, in the wake of modernization, left their rural homes to work in the cities, *furusato* or old country home lives on in the memory as a place where parents, grandparents and other family members live, a place of childhood and a place of beautiful memories of days gone by. For those dreaming of returning to their rural homes in triumph after succeeding in the city, the city represented a battleground where difficult obstacles had to be overcome. In contrast, that rural home was thought of as the place where parents and relatives looked on from afar and waited for their return. It must be said, however, there are also many Japanese who reject their old country home because it brings to mind a pre-modern, inbred society with a unique set of social complications and difficulties. For such people, the song *Furusato* is something to be repudiated. We can correspondingly understand that the *furusato* image inspires both love and hatred.

Since the disaster, furusato has gradually become the symbol of lost nature and of lost homes and families. When people talk about the catastrophe, they talk about the lost natural environment of Tohoku and conjure up images of the typical farm, fishing and mountain villages of the Tohoku (northeastern) area of Japan, which are evoked by the words homes and families. Against the background of conjuring up these images of typical farms and now with mountain village recovery projects, the most challenging problem confronting city planners becomes how to somehow interrupt the intimate attachment each person who lived in the affected area feels for his or her lost old country home and lost land.

The sense of loss experienced by disaster victims is not restricted to the concrete loss of family members and houses. Those victims have also lost the time and space that they shared with their families. They have even lost their photographs and, it is safe to say, all evidence of themselves as human beings who lived with their families. This is an important element that constitutes "being alive."

It seems reasonable to say that the loss experienced by disaster victims who hum the song *Furusato* to themselves is in fact *oikos*. On the other hand, we should also remember that awareness of the loss of home is not restricted to disaster victims. We humans carry within our bodies the memory of happiness associated with our younger years, especially the peace and calm we experienced under the protection of our parents. That unspoiled scenery of happy experience is called to our minds whenever we confront hardship or sorrow later in our lives. Sometimes, our longing for that unspoiled scenery helps us extinguish the feeling of alienation that arises between ourselves and our current condition. At other times, this might serve to heighten our recognition of our discontent or unhappiness with the current situation. The feeling of well-being and happiness we felt in the past becomes a nucleus of ourselves as a primal experience. Though not sharply defined, this provides us with a reliable fixed point that we continue to recognize within us as a "feeling." Even amid the uncertain and unceasing movement that characterizes human life, we have a place, a fixed point that enables us to keep our balance and allows us to perceive the most stable fixed point from moment to moment. That place can be called *oikos*. The desire to recover the lost *oikos* that represents the original point of our childhood peace and happiness is precisely what defines our humanity, our "being alive" as flesh-and-blood people who live in a corporeal reality.

Oikos is repeatedly recollected as a longing for home or through the complex interplay of emotions that combine love and hate. Through repeated telling, we weave for ourselves a story of our old country home. Throughout world history, the keyword "home" has played the role of giving people courage and hope in the recovery period after a war or a disaster, appearing in literature, film, and other media.

For examples, we can look at Germany in the aftermaths of both World War I and World War II, when cultural and artistic movements arose that focused on the recovery of home. For Germans completely exhausted by defeat, the word *Heimat* (which denotes a love of homeland) was a keyword for reconfirming identity and inspiring the courage and hope needed to rebuild their country. Of course, the word was also deployed during the wars as a tool for raising national prestige. In the 1920s, the value of home was rediscovered, and *Heimatkunde* (local history) became part of the school curriculum and an important means of encouraging an interest in the German state and a love of country among children. Similarly, the *Heimatfilm* (homeland films) made in the 1950s, which are set in the rich natural beauty of the mountain villages in southern Germany during the postwar reconstruction period, were part of a cultural policy designed to help give the German people the courage and strength to recover.

It cannot be denied that states have applied images of home to enhance national prestige and establish ethnic identity. It is also true that exiles who have lost access to their countries are different from disaster victims whose mental state vis-à-vis home reflects a loss of familiar people and things as well as the communities and families to which they once belonged. However, there is a point of commonality between exiles who have been forced to leave their countries and the lives they lived there and disaster victims who are still in the geographical location of their homes but have lost those homes. In both cases, the people in question must reconcile their former identities, which have been implanted in the ground with their identities as an "image of home." In other words, both are searching for their lost *oikos*.

The search for lost *oikos* is not limited to exiles and refugees. *Oikos* helps people secure their own identity and gives them a place to come from and return to. As something that once was and is no more, it becomes an object of aspiration to all people. It might be a longing for the time and space of childhood. Or it might be an abstract but fixed point that people want to always have established for themselves as a place to be. The question is, how can we gain this feeling of living within our bodies in the sense of "being alive=dwelling= living", this feeling of existing here in this place, this feeling of fitting perfectly where we find ourselves?

Human beings in various cultures and societies have acquired wisdom concerning this question through ceremony, custom and mimetic learning. We would do well to explore that accumulated knowledge and apply it to our modern age. In the remainder of this paper, I'll focus on the wisdom found in Japanese-style thinking and, while referencing the recent disaster, explore the knowledge that might be useful for recovering lost *oikos* in relation to the Japanese sense of happiness. I believe that it is incumbent upon historical-cultural anthropology to clarify the human condition as an axis of the desire for lost oikos and to achieve this by exploring the diversity of wisdom and knowledge acquired by various cultures and societies alongside identifying their points of commonality.

4 *Oikos* in Relation to *topos* (*basho*, 場所) and Body (*mi*, 身)

Oikos is embedded in *topos* and body. *Topos* differs from the concept of space found in Euclidean geometry or Newtonian physical laws, where it is abstract and homogeneous. Originating with Aristotle, *topos* existed before the emergence of the modern concept of space.

According to Aristotle, *topos* means a symbolic and bodily space (Aristotle 1960); applied, for example, to refer to laws, where it is abstract and homogeneous. A symbolic space contains a fulfilling meaning and significant direction like a sacred space or a mythical space, which are distinguishable from a secular space. Sacred space as a whole takes on the character of the universe and often appears in a form that resembles the cosmos. In ancient times, cities or houses were built with an awareness of the cosmos or the universe. A *genius loci* appears as the peculiar atmosphere of a place.

A space with a bodily nature means not only the space that is internalized in the body but also the body itself as a substratum, a space of spirit or the conscious self. On the one hand, this bodily, internalized space is articulated by the activating or territorialized body. But, on the other hand, the body as a substratum of spirit or of conscious self means not the physical body (which Aristotle viewed as the prison of the spirit and Descartes viewed as a substantial entity that was distinguished from the spirit), but the existential body, which is the horizon and the foundation of the conscious self.

Space as the foundation of the human being or of being itself has long been sought in the sphere of the subject and of existence. For Aristotle, it meant being a sublime, God-like being; for Descartes, it meant being a self-subject. In contrast to this western philosophical framework, Kitaro Nishida identified basho (place) as the foundation of being, defining it as both object and nothingness. This is his "logic of place" (Heisig 2001, 72–73; Elberfeld 1999, 103–104).

Nishida accepted the subject-object-frame of German Idealism but tried to reinterpret the phenomenon as occurring in a place. He believed that all phenomena emerge through the interactive relationship between the human being as acting subject and by the place as passive object. He tried to overcome the dualities of subject/object and active/passive through the application of Zen Buddhism and other traditional Eastern thought systems.

He believed that of all phenomena emerging through this, it becomes important to analyze not the being itself, but the "something" between human beings, which defines the whole condition of the place as the "in-betweeness" or better to say the blank or thread found between human beings. This "something" activates the being itself.

Oikos in the Japanese context is a connector binding the macrocosm (the universe, the world, nature) and the microcosm (the bodies of individuals, human beings). The Japanese word "mi" is slightly different from "body" in Western mind-body dualism (Suzuki 2010.) A mi (body) exists in relation to others and can probably also be spoken of as forming, as it were, a network of relationships. If an awareness of one's own mi can be called a physical awareness then the mi that is understood by this physical awareness becomes subtly different in its relationship with others; even so, it can probably be said to possess what might be called a nucleus that remains unchanged no matter whom it is with or when it is with them.

As is well known, the term "body" acquired a new meaning with the development of conceptual methods based on the physical sciences in the eighteenth-century West. It no longer simply signified a flesh-and-blood body as it had before but was invested with the meaning of "material object" in the language of physics. A modern model of human behavior was also devised in which the human body, conceived as a material object in the scientific sense, lives and acts in a uniform spatio-temporal grid. The German word *Körper* signifies "body" in the modern sense of "material object" and is distinguished from *Leib*, the flesh relative to the soul in Christianity (Feher 1989; Suzuki 2010).

Thus, a human model was worked out based on this modern view of the body – that of the human being as a homogeneous unit capable of operating purposely and efficiently if it so wills. In the modern system, which constitutes a uniform model in which human beings are considered constituent parts that make up a state (that is, modern citizens), the norm on which human beings were modeled was the life cycle of an adult, especially an adult male. The labor market regards a human being as a uniform source of labor power, as a working person or as a body that works. Based on this modern view of the body, the present educational system can be said to function in a framework where human abilities are regarded as social resources.

In this modern view of the body, the body is seen as a closed system that generates the power of self-preservation through the relationship between stimulus and response, action and reaction. But who activated this system called the body in the first place? Before the modern period that higher being was thought to be God. Now, in the modern worldview, the stage in which this system was first set in motion cannot be dealt with in a modern scientific explanatory framework, so the question is, as it were, put in parentheses and ignored. Though inquiry into the existence of a higher being is beyond the scope of science, still, human beings are alive; indeed something has caused them to be alive. Researchers have been aware of the difficulty of understanding the human essence in terms of modern conceptual methods alone. Since the eighteenth century, they have constantly warned of the need to consider not only a body's active side but also its passive side. This has led to observations of the body, its feelings and emotions which might be called the irrational side of human beings and the formation of several schools of thought that have survived down to the present day.

5 In Search of the Lost *oikos* in the Style of Japanese Thinking

After 11 March 2011, many foreign media gave full coverage to the endurance and self-control of Japanese people. On the one hand, they appreciated the organized and self-controlled behavior of the Japanese after the earthquake and tsunami. On the other hand, they wondered why the Japanese did not criticize the regulatory policies of the Japanese government and TEPCO (Tokyo Electric Power Company, owning the stricken plant). Foreign media may feel more frustration concerning the lack of credible information. From the Western view, it is nothing less than the patriarchal attitudes of Japanese people. Is it resignation? Is it spiritual awakening? Is it a giving up of positive meaning or an adjustment? Whatever it might be, I think it does coincide with the Japanese mentality. I would like to review the background of this mentality as it relates to the pursuit of peace and happiness.

The media gave full emphasis on the stance and self-control of Japanese people. They appreciated the organized and self-controlled behavior of the Japanese in accordance with *wazawai*, 災い; *waza* means the work of God and *wai* means the effect of his work. The first part of the word is written with the Chinese character for disaster (災), which means damage from heaven, such as fire. This is combined with a Japanese hiragana character to form *wazawai* (災い). Compared with the Chinese concept of disaster, *wazawai* means providence or the work of God. It is not, however, the work of the Devil, who brings human beings something bad. This is the distinction between the Japanese concept of God and the Chinese one. In Japanese, God is given absolute credit for his work. God has the best interests of human beings in mind and deeply considers the best way for us to know the meaning of our lives.

The opposite of *wazawai* is *saiwai* (幸い); this is also God's work, but characterized by *sai* or goodness. Both words appear in the Japanese proverb "*Wazawai ha saiwai no tane*", which means: "Disaster is a seed of happiness." It is important to point out that for the Japanese, both goodness and badness exist in the hands of God and their contradiction channels a situation comparable to an on and off switch. Disaster and happiness are all one and change their positions in relation to each other depending on the context. The result is a dynamism that turns unhappiness into happiness and vice-versa.

I would like to quote a student named Yuuta Kajiwara, who wrote the following message after the triple disaster that hit north eastern Japan in 2011. Yuuta is a graduate of Hashigami Junior High School of Kesennuma City in Miyagi Prefecture. Kesennuma was also hit hard by the disaster. The excerpt is part of the speech he gave at his graduation ceremony, which was delayed because of the disaster.

The disaster took by force many important lives and things from us. Its damage is so brutal that I consider it an ordeal or trial by heaven. It pains me. I am bitterly disappointed. My alarm clock stopped at 14:46, the moment the disaster struck. But time flows steadily. We who survive have to live our lives honestly and strongly with compassion in a positive light. We have learned the importance of life from this enormous disaster. But the price for this lesson was too high. Nevertheless, it is our mission to live without blaming heaven, to live by accepting our fate with a spirit of helpfulness in a difficult situation.

Here we can recognize the typical style of Japanese thinking concerning disaster and happiness. However, is this style really only found in Japanese culture? How much is it influenced by the cultural background? The difference between pre-modern and modern cultures sometimes appears to be the same as the difference between Western and Eastern cultures.

Happiness depicted in a Japanese proverb: *Kafuku ha azanaeru nawa no gotoshi*, 禍福はあざなえる縄のごとし (Good and bad are twined together like a braided rope). Because good luck and bad luck, like a braided rope, come in turns in human life. This proverb contains the moral that life should be lived without despairing or becoming overly depressed by misfortune and without becoming excessively elated when blessed with happiness or good fortune. It describes an attitude toward life that sees human existence as having ups and downs. At the same time, it might also be said to contain lessons for what might be called a philosophy of life and one in which a person reflects on his or her attitude toward his/her own life based on the various events that occur to other people. By comparing one's own life to that of others, one comes to have a sense of awe for the mysteries of human existence and for the forces that transcend human beings and, at a deeper, hidden level, direct their actions. It also gives one the opportunity to gain an insight into one's own life mimetically, as it were, by taking the bird's eye view of someone else.

This calls to mind another saying in Japanese: *taru wo shiru*, 足るを知る (learn to be content), which is a Buddhist teaching derived from Zen. A person who is not able to understand what constitutes a state of contentment for him/herself is always craving something and suffers from that craving. To understand what personal contentment is, a person must know his/her own measure – in Japanese the expression is *mi no take*. It is important to know one's own size, to know what is truly one's own just size.

Let us consider happiness from the perspective of a meaningful life or the joy one feels in being alive (Wulf et al. 2011). The happiness one feels when one deems life to have meaning in this sense can be said to occur when a person encounters a situation in which he or she is inspired by or responds to some other person or thing. When one has the good fortune to meet a person or thing that stirs one's soul, one is grateful for that encounter. The feeling of gratitude for that encounter reaches the point where one comes to believe in a connectedness to the world.

Such an encounter changes the perspective on a meaningful life, or on the joy one feels in being alive (Wulf et al. 2011) The happiness one feels when one deems life to have meaning in this sense can be the process that leads from chance encounter to gratitude to belief to conversion is not however necessarily an easy or placid one. Indeed, it may bring a greater degree of anxiety in the sense that one is placing oneself in the very midst of change.

Happiness might be described as the courage to withstand the uncertainty one feels in the midst of this process. The greater the anxiety and uncertainty, the greater the joy engendered by the encounter and the feeling of gratitude for it. Human beings by nature are supposed to crave stability. To have the courage to accept and come through unsettled conditions and to feel happy doing so, necessarily presupposes a desire for a deep spiritual encounter with some person or thing and a feeling of gratitude for that encounter.

The relationship between risk and happiness will need to be tested from many angles in terms of the courage of people facing uncertainty and their attitudes toward it. However, even more to the point when people have the courage to seek encounters under changing and transformative circumstances, happiness has probably already ceased to be an object of active pursuit for them. It might be said that a person is most happy when in a state of not desiring happiness. This may well be one of the challenges of an international comparative study of happiness – determining how it is possible, through a modern scientific approach, to account for this paradox of happiness.

It is important to consider what kind of wisdom is necessary for human beings to live with the risk of disaster. This might be a key to the pursuit of happiness. People have come to seek not only material affluence but spiritual richness as well. We need to know what wealth is in the truest sense of the word.

The integration of economy and ecology with the ancient Greek concept of *oikos*, I suspect, can lead to a breakthrough in the development of a new intelligence for future survival strategies. For human beings, living in *oikos* means living in relation to the world around us from the perspective of mind and body. When it comes to both economics and ethics, is it not oikos that will help us to think together about how to secure a place where we can exist as living beings and discover once again the meaning of life?

References

Arendt, Hannah. *The Human Condition*. Chicago and London: University of Chicago Press, 1958.
Arendt, Hannah. *The Life of the Mind*. London: Secker & Warburg, 1978.

Aristotle. *Posterior Analytics. Topica*. Trans. Hugh Tredennick and E.S. Forster. Cambridge: Harvard University Press, 1960.

Elberfeld, Rolf. *Kitaro Nishida (1870–1945). Das Verstehen der Kulturen. Moderne japanische Philosophie und die Frage nach der Interkulturalität*. Amsterdam: Rodopi B.V., 1999.

Feher, Michel, Ramona Naddaff, and Nadia Tazi, eds. *Fragments for a History of the Human Body*. New York: Zone, 1989.

Finley, Moses I. *The Ancient Economy*. Berkeley: University of California Press, 1999.

Heisig, James W. *Philosophers of Nothingness*. Honolulu: University of Hawaii Press, 2001.

Ingold, Tim. *Being Alive. Essays on Movement, Knowledge and Description*. London and New York: Routledge, 2011.

Le Goff, Jacques. *Your Money or Your Life: Economy and Religion in the Middle Ages*. Trans. Patricia M. Ranum. New York: Zone Books, 1988.

Nishitani, Keiji. *Nishida Kitaro*. Trans. Yamamoto Seisaku and James W. Heisig. Berkeley: University of California Press, 1991.

Schumpeter, Joseph A. *A History of Economic Analysis*. Oxford: Oxford University Press, 1954.

Suzuki, Shoko. "Sein und Werden – Poiesis in Gesten." In *Gesten*, eds. Christoph Wulf and Erika Fischer-Lichte, 116–122. Munich: Wilhelm Fink, 2010.

Tuan, Yi-Fu. *Topophilia: A Study of Environmental Perception, Attitudes, and Values*. New Jersey: Prentiss Hall, 1974.

Vernant, Jean-Pierre. *Myth and Thought among the Greeks*. Trans. Janet Lloyd and Jeff Fort. London: Routledge and Kegan Paul, 1983.

Wulf, Christoph, Shoko Suzuki, Jörg Zirfas, Ingrid Kellermann, Yoshitaka Inoue, Fumio Ono and Nanae Takenaka. *Das Glück der Familie. Ethnographische Studien in Deutschland und Japan*. Wiesbaden: Springer, 2011.

Part II: **A Casuistry of Disaster**

Elsa Alves
The Specter of Chernobyl: An Ontology of Risk

> Knowledge has in us been transformed into a passion which shrinks at no sacrifice and at bottom fears nothing but its own extinction; we believe in all honesty that all mankind must believe itself more exalted and comforted under the compulsion and suffering of this passion than it did formerly, when envy of the coarser contentment that follows in the train of barbarism had not yet been overcome. [...] Yes, we hate barbarism – we would all prefer the destruction of mankind to a regression of knowledge! And finally: if mankind does not perish of a passion it will perish of a weakness: which do you prefer? This is the main question. Do we desire for mankind an end in fire and light or one in the sand? (Friedrich Nietzsche, *Daybreak*, § 429).

Nietzsche in this text announces the birth of a "new passion", the passion for knowledge, against the "old passion", the passion of Christ, thus aiming to triumph over morality. However, this new passion is also seen as an imminent tragedy, self-sacrificing humanity to the scientific truth, which, apart from its noble intentions, proves as illusive a goal as the moral truth and therefore constitutes another nihilistic trap that can only lead to a ruinous ending. In another aphorism in the same book (§ 327), Nietzsche even compares the lover of knowledge to the tragic character of Don Juan, in his insatiable and vertiginous chase only for the pleasure of the hunt at the end of which "Hell is the last knowledge that *seduces* him". The eschatological finitude enounced by Nietzsche in his rhetorical question – "do you desire for mankind an end in fire and light or one in the sand" – became the horizon – ontological, cultural and socio-political – that has now been called *risk*. In other words, when God dies, risks necessarily begin to emerge.

The Chernobyl disaster was perhaps the most emblematic historical example to illustrate the "sacrifice", on the one hand, and the "absence of fear", on the other, that Nietzsche referred to regarding this so-called "new passion". They are both present, for instance, in the "character" of the 800,000 liquidators who volunteered to minimize the damages, nonetheless fighting the Promethean fire, and in all the inhabitants of the contaminated zone. That heroism, vested by a large dimension of pain and the necessary denial (or ignorance) of risk, was perhaps never so well personified. However, by being an ongoing nuclear catastrophe, Chernobyl is not only a historical landmark of that risk-creating "new passion" but also carries a radically different nature concerning risk, defined by its latency and its temporal and spatial omnipresence. From the moment the "risk society" takes hold, the catastrophic event, which is by definition an exceptional and singular event out of the historical *continuum*, becomes the model, whether ontological or cultural, of a new state of things. I thus discuss the Chernobyl disaster understood as a universal accident and, consequently, as a universal threat to our imagination.

1

Accordingly, the understanding of risk I wish to undertake is much wider than the strict sense of the risk management lexicon and, to a large extent, all of the techniques concerning the governmentality of catastrophes and other types of risks in contemporary society. Risks indeed establish themselves as relations of power-knowledge in late modernity for they are an abstract product of expertise upon which government is based. Nevertheless, beyond those strategies, the Foucaultian perspective is not concerned with the nature of risk. On the other hand, in a market economy, risks may be perceived as opportunities rather than hazards to be avoided and may indeed have been increasing at the pace of the pursuit of profit according to a capitalist logic of "creative destruction". However, when dangers reveal themselves, threatening both the exploiter and the exploited, this hypothesis also reveals itself insufficient to explain the phenomenon. Drawing on Ulrich Beck, the broad meaning of risk I wish to advocate presupposes that hazards objectively exist, which are or are not subsequently translated into the probabilistic rationality of risks. As Beck (1992, 44) says, "what escapes perceptibility no longer coincides with the unreal". Thus, risks exist with or without our perception of them even though their symptoms always remain perceptible. The "risk society" is, therefore, as Beck argues (44), the rise of a Platonic world breaking down the Aristotelian in which reality, at least the most determinant reality for us, falls beyond our senses. Likewise, Jacques Derrida (1984), referring to the issue of nuclear war, told us that notwithstanding our modern "rhetorical condition" we could not entirely dismiss the possibility of total destruction. Similarly, Jean-Luc Nancy (2012) states in his recent essay on the Fukushima disaster that the possibility of absolute devastation is a kind of singular decision that overcomes all deliberation, i.e. all the techniques of resilience used by risk management to render hazards calculable and governable.

Thus, what I mean by an ontology of risk involves how its status, apart from its constructed compound, cannot ultimately be reduced to a metaphor. Even if we cannot trace, for instance, a relationship between risks, scientific knowledge, media staging, individual perception and social representation we must, nevertheless, assume that there is a risk; the *virtuality* of a risk is thus *real*. On the other hand, I choose to use the term risk and not "uncertainty" (Beck 1992), or "vulnerability" (Lemarchand 2002), among others, because I believe this kind of distinction carries little use even in heuristic terms when one aims to grasp the ontological condition of being-in-threat or being-at-risk since, in the first place, all kinds of critical approaches always arise *within* the normative assertion of risk; and secondly, from the moment that highly improbable events appear to us as probable, that is, when "the calculation of risk collapses" (Beck 1992, 22), allegedly inaugurating the "risk society", the boundary between empirical risks and generalized threats begins to blur and hence becoming commutable terms.

In order to clarify this point, we return to Nietzsche's quotation which gives us another important insight about the nature of risk moreover conveyed in Beck's notion of "reflexive modernity": the imminence of disaster stems from the old illusion of modernity, namely, techno-scientific progress. This "meta-narrative" that was once claimed as lost appears to be, on the contrary, more inexorable when risks began being enunciated as *risks*. As Martin Heidegger said decades before Beck: "All attempts to reckon existing reality [...] in terms of decline and loss, in terms of fate, catastrophe, and destruction, are merely technological behavior" (Heidegger 1977, 48). The very act of diagnosing in any systematic way the potential dangers, and nominating them as such in order to solve them, is in itself totally technocratic as the more risks are identified, the more man wants to be in control of technology. The narrative of risk, i.e., the anticipation of catastrophe and the risk society organized around this anticipation of catastrophes thus belongs to a radicalized narrative of progress set in a vicious circle. Correspondingly, according to the "precautionary principle" proposed by François Ewald (2009), which is theoretically close to the "heuristics of fear" of Hans Jonas (1992), we produce ever more risks or make them visible (which is an inclusive disjunction for both statements hold true) and, for such reason, we must take into consideration the most improbable hypothesis. This is so because man is the potential author of the disasters, the only entity responsible for the increasing of endogenous risks arising out of modernization. This also represents the reason that a "pedagogy of disaster" (cf. Lemarchand 2011; Debray 2011), that is, the possibility of learning from a disaster, seems likely to be impossible right from the outset. Consequently, the idea of a "reflexive modernity" that would ultimately be able to transform itself through self-criticism becomes perhaps too optimistic.

On the other hand, precisely because risks are no longer foreseeable, there is a generalization and hyperbolization of doubt, or, in François Ewald's words, there is a return of Descartes' malicious demon standing in the way of truth

(Ewald 2009). Before the incompetence of science, one can no longer separate belief from knowledge, *doxa* from *episteme* (Derrida 1984). Thus truth does seem to be regaining, as mentioned above, a metaphysical dimension beyond our Kantian "condition of possibility of knowledge", since what threatens us most has become invisible. This epistemological crisis, in turn, easily leads to the creation of imaginary causes, to the dissemination of a culture of fear and fiction in society, presented and intensified by the media (cf. Weart 1988; Walters 2008). Consequently, our risk susceptibility and risk perception change both with the revealing of new risks by science and with real events dramatized by the media and art. What happens in fact is a dialectical relationship between the social risk discourse, individual perceptions and real threats. The reality of a risk, thus, always beckons as a specter.

Therefore, the advent of science and technology did not completely abolish the faith in spirits and other ghosts. Technology itself can actually be seen as a secondary sacralization, a substitution for the previously killed God (Dupuy 2006). Furthermore, civil and military nuclear technology was determinant to this new enchantment and everyday speculation as the danger of radioactivity completely escapes our senses. As the historian Galia Ackerman eloquently puts it (cf. Dupuy 2006, 48): before the contaminated zone of Chernobyl we are more touched by metaphysics than by physics.

2

The Chernobyl disaster is many things, among them a myth, or rather a series of myths. The most recurrent trope, however, that tries to ascribe a sense to the absurd, is the apocalypse, which has so ironically been *revealed* (the original meaning of the word apocalypse) a priori in its name as the meaning of the Ukrainian word "Chernobyl" is wormwood, the name of the star that, according to John of Patmos, would forever poison the waters. But embedded in these stories, there is the "material" Chernobyl that has to be understood as a cultural phenomenon as well as an ontological model of contemporary risk.

As I said above, the risk society is determined by the invisibility or the virtuality of the real. The Chernobyl disaster is to be read, above all, as an invisible hazard and, for this reason, as a new way of perceiving the world. In Peter Sloterdijk's words (2009), with the nuclear, there is also a catastrophe of the phenomenal (and not only the catastrophe as a phenomenon): "the need to perceive the imperceptible hung over [us] like a new law formed as a threat". On accepting technology as a sort of an extension of man, that is, man's imagination transforming the world in his own image, with nuclear technology and specifically with Chernobyl, the first big techno-scientific accident, we may now instead observe the technology surprising man and challenging his/her own imagination in an unprecedented way. Therefore, nuclear catastrophe poses specific problems and limits not only our perception theory but also our imagination.

When the fourth reactor of "Lenin's" nuclear power plant in Chernobyl blew up on the night of 26 April 1986, large amounts of radioactivity were released into the atmosphere and first registered two days later by workers at the "Forsmark" nuclear power plant in Sweden, over one thousand kilometers away from Chernobyl. This first alarm warning became the indication that Chernobyl was not just an incident inside the borders of the then Soviet Union but, instead, a global accident. By global, this incorporates not only the media impact of the disaster but also the materiality or reality of the radioactive contamination. After that firsthand notification and official confirmation by the Soviet Government, the circumstances of the accident and that invisible threat, often mystified as a "cloud", were copiously interpreted both by the media and by scientific experts. Subsequently, the scientific accounts, whether from the International Agency for the Atomic Energy or other agencies, have stated disparate conclusions about the facts concerning the accident until now, 26 years after the accident. The most controversial facet relates to the number of victims with the total of deaths ranging from 4,000 (according to the Chernobyl Forum Report published in 2005) to 965,000 (according to the New York Academy of Sciences report published in 2009). The reasons for this disparity underline the epistemological crisis I referred to above related to this new invisible and virtual (although real) threat. Determining a causal relation between certain amounts of radiation exposure and objective diseases has proven highly complex and polemic. However, without scientific measurement instruments, radiation would not even have been perceived and accepted. Faced with radiation, only science can serve as a means of mediation; in the same way only that communicated through the media becomes *real*, but only to a certain degree. Even if the senses are incapable of seeing the danger and science is somehow incapable of accurately determining the risk, the cells will nevertheless react to it and, later on, the symptoms will confirm that reality manifested in an obscene way: either in congenital deformities or children dying of cancer. Thus, the risk has a radically new dimension of latency (*idem*); as Sloterdijk says referring to Heidegger's concept of homelessness: man's condition is homeless now, not only because his environment is being destroyed, but also and mainly because there are no possible protective niches against the latency of risk (*idem*).

The constant need to update the number of deaths also tells us that the Chernobyl disaster was a singular event in the past, extended into an endless future and thereby occupying an indeterminate space. Consequently, Chernobyl created new temporal and spatial dimensions by the silent poisoning of humans and the environment. French sociologist Frédérik Lemarchand terms this a new model of epidemic catastrophe following Jean Baudrillard's (1993) notion of the "transparency of evil". The old culture of epidemics, based on the logic of confinement, is now obsolete. Formerly, physicians wanted to know where to bury the corpses and keep patients in compulsory isolation. On the contrary, the "sarcophagus", the concrete envelope surrounding Chernobyl's reactor, was designed to contain the radioactivity and, paradoxically, it is outside that sphere that people find the "shelter" wherein they hope to be safe (Lemarchand 2002).

Radioactive contamination, instead, develops in a dialectical relationship between long-term temporality and urgency, contamination at the global scale and contamination in specific zones. However, even within specific areas, the exclusion zones, there are the so-called "hot spots" of radiation, which reveal differences in the incidence of radioactivity to an almost infinitesimal degree. Moreover, it actually becomes impossible to state that outside the delimited zone we are safe or only inside it are we in danger. Equally impossible is any statement claiming only some areas are contaminated or that everything is equally contaminated.

On the other hand, Chernobyl also demonstrated how radiation could no longer be understood from a linear causality (Adam 1998). Given that effects can be displaced to varying degrees both in space and time, extended from now to subsequent generations, without any possible reversibility, that theoretical assumption must be reconsidered. The (im)possibility of reversibility, in fact, becomes a very interesting case in Chernobyl as people can watch the plants grow and the animals reproduce in apparent normality (Mycio 2005). However, this also denies the real risk and hides the damned future for generations of human and animal beings to come. If risk is deferred in time and dispersed in space, in a unique progression, it escapes not only our perception but also our science, at least its classical way of measuring and tracking causalities. The Chernobyl disaster has therefore shown that radiation represents an incommensurable hazard and thereby calling into question the scientific assumption that risks are knowable, measurable and predictable (cf. Beck 1992; Adam 1998).

3

The Chernobyl disaster is not only a disastrous outcome of Nietzsche's "passion for knowledge" but also an ironic consequence of the "passion for real" as coined by Alain Badiou, referring to the twentieth century's driving force, that is, the nihilistic pursuit for destruction, for the nothing, as the only thing that remains real for "it is impossible to seem to die" (Badiou 2007). However, as Slavoj Žižek (2002) points out, regarding the September 11 attacks, the "passion for real" instead ends up in spectacularization. The tragic real, because of its excessive character, is only sustainable when one fictionalizes it; in short, it can only appear to us as a specter. This is undoubtedly true as we mentioned above regarding the threat from radioactivity even if taking into account that this might be overdramatized by the media and not exactly measurable by science, the fact remains that we only know about its existence through media and science mediation. Hence, this only exists to us as a specter, that is, in its dialectical presence and non-presence to make recourse to Derrida's concept of specter's ontology or "hauntology" (Derrida 2006).

Simultaneously, however, when the threat is invisible and potentially omnipresent, how can it be subsumed by any system of representation or analogy and how can it be processed by experience? In a society where evil becomes more and more epidemic, when it is impossible to perceive what is poisoning us, to what extent and for how long and when impossible to determine with any certainty the cause of certain symptoms, "metaphor is no longer possible". This is Baudrillard's thesis on a new way of evil (2007), following on from Deleuze and Guattari's theory of becoming that I seek to apply here to the paradigmatic case of nuclear risk. When radiation penetrates and modifies our genes, there is no more "like-something", this amounts to the abolition of all comparison or metaphor and "all that consists is Real" (Deleuze and Guattari 2004). What happens, instead, is a metamorphosis or a transmutation, a becoming-something-else, malformed or not identifiable, virulently undermining not only the physical body but also the symbolic order.

This is the corollary of my hypothesis: in the face of the Fukushima disaster, the specter of Chernobyl warns us that the more a catastrophe is plausible or even real, the less it can be imagined, thus resisting the symbolic order. In her essay entitled *Depraved Perspective* (1991), Annie Le Brun had already discussed this problem regarding the nuclear situation, distinguishing it from other types of catastrophes. She argues that since the eighteenth century, the meaning of catastrophe had been the Greek one, that is, an event that would change the order of things and inaugurate a new world. However, with the nuclear catastrophe and/or its imminent possibility, the old dialectic between real catastrophe and imaginary catastrophe was abolished. If the catastrophe is no longer an exceptional event and if it is a human deed, no longer caused by nature or God, then it escapes from our imaginary to become our real threat. Until the nuclear condition, the catastrophe would serve as a symbolic construct around which humanity would search to define itself, under the sign of accident, chance or fate. However, from the moment that the accident becomes "prophetic", the term deployed by Virilio (Virilio and Alexievitch 2006) to describe the Chernobyl disaster, that is, when an accident has a permanent incidence, then the catastrophe loses its pivotal mission of confronting the human with the inhuman, i.e., of reaffirming identity through its contrary.

When we are constantly living in a "state of exception" (Agamben 2005), that is, in permanent alert, guided by a "precautionary principle" (Ewald 2009), the censure over the original feeling of catastrophe occurs or, in other words, the catastrophe loses its singular character and becomes a banal event. Living in a permanent catastrophe, as in the case of Chernobyl, means, therefore, trivializing it, making it natural, facing the Eternal Return of the sameness (without the "Zarathustrian" joy). As Jean-Luc Nancy (2012) argues, with nuclear technology, the scope for damage becomes of such a magnitude that it no longer proves possible to calculate in terms of relative forces because a nuclear impact engraves itself into human and natural tissues, hence cancelling, on the one hand, the old dichotomy between nature and culture and, on the other hand, the possibility of conceiving force strategies: "There's no more point in representations like 'David and Goliath', 'Ulysses and the Cyclops', or 'Zaitochi, the blind samurai'" (Nancy 2012, 41). In other words, there are no more tensions, dialectics or differences, only equivalence and sameness. Jacques Derrida (1984) has already stated the same argument when discussing the possibility of a nuclear war: on the one hand, perceived as just speculation and not considered as serious or even real; on the other hand, the nuclear war as the only possible referent, the absolute horizon since it presupposes the total destruction of the archive and thus, also that of metaphor. An individual death or the destruction of part of a community can become the object of mourning and memory but the spectral presence of a nuclear catastrophe threatens not only the real human being, but also his symbolic and imaginary dimension. Thus, the only referent that is absolutely real is, paradoxically, the specter of nuclear disaster.

Chernobyl as the first big techno-scientific catastrophe, both as a potential and actual catastrophe, is a spectral reality, invading our imagination and defying our symbolic world while rendering necessary what is contingent by definition, substantializing the accident and imposing "live the future already in the present" on us (Lemarchand 2006b). The Chernobyl disaster is and represents the emergence of a new spatial and temporal culture in which scientific knowledge plunges into deep doubt and being-at-risk has become our new condition, as we are trying to face the invisible, global and imminent threat.

References

Adam, Barbara. "Radiated Identities: Invisibility, Latency, Symptoms – The Case of Chernobyl." In B. Adam, *Timescapes of Modernity: The Environment and Invisible Hazards*, 193–210. London: Routledge, 1998.
Badiou, Alain. "The Passion for the Real and the Montage of Semblance." In A. Badiou, *The Century*, 48–57. Polity Press: Cambridge, 2007.
Baudrillard, Jean. *The Transparency of Evil*. London: Verso, 1993.
Baudrillard, Jean. *Forget Foucault*. Los Angeles: Semiotexte, 2007.
Beck, Ulrich. *Risk Society: Towards a New Modernity*. London: Sage, 1992.
Debray, Régis. *Du bon usage des catastrophes*. Paris: Gallimard, 2011.
Deleuze, Gilles, and Felix Guattari. *A Thousand Plateaus: Capitalism and Schizophrenia*. London: Continuum, 2004.
Derrida, Jacques. "No Apocalypse, Not Now." *Diacritics* 14, no. 2 (1984): 20–31.
Derrida, Jacques. *Specters of Marx: the State of Debt, the Work of Mourning and the New International*. London: Routledge, 2006.
Dupuy, Jean-Pierre. *Retour de Tchernobyl: journal d'un homme en colère*. Paris: Éditions du Seuil, 2006.
Ewald, François. *Le principe de précaution*. Paris: PUF, 2009.
Heidegger, Martin. *The Question Concerning Technology and Other Essays*. New York: Harper Torchbooks, 1977.
Jonas, Hans. *Le principe de la responsabilité: Une éthique pour la civilisation technologique*. Paris: Cerf, 1992.
Le Brun, Annie. *Perspective dépravée: entre catastrophe réelle et catastrophe imaginaire*. Bruxelles: La Lettre Volée, 1991.
Lemarchand, Frédérik. *La vie contaminée. Éléments pour une socio-anthropologie des sociétés épidémiques*. Paris: Harmattan, 2002.
Lemarchand, Frédérik. "Post-scriptum: mythes du monde d'après l'apocalypse." In *Les silences de Tchernobyl: l'avenir contaminé*, eds. Galia Ackermann, Guillaume Grandazzi and Frédérik Lemarchand, 211–225. Paris: Editions Autrement, 2006a.
Lemarchand, Frédérik. "Le futur pour mémoire." In *Les silences de Tchernobyl: l'avenir contaminé*, eds. Galia Ackermann, Guillaume Grandazzi and Frédérik Lemarchand, 261–277. Paris: Editions Autrement, 2006b.
Mycio, Mary. *Wormwood Forest: A Natural History of Chernobyl*. Washington, D.C.: Joseph Henry Press, 2005.
Nancy, Jean-Luc. *L'équivalence des catastrophes (Après Fukushima)*. Paris: Galilée, 2012.
Nietzsche, Friedrich. *Daybreak: Thoughts on the Prejudices of Morality*. Cambridge: Cambridge University Press, 2003.
Sloterdijk, Peter. *Terror From the Air*. Cambridge: MIT Press, 2009.

Virilio, Paul, and Svetlana Alexievitch. "Êtres temoins de cet accident du temps." In *Les silences de Tchernobyl: l'avenir contaminé*, eds. Galia Ackermann, Guillaume Grandazzi and Frédérik Lemarchand, 228–236. Paris: Editions Autrement, 2006.

Walter, François. *Catastrophes: une histoire culturelle XVIᵉ–XXIᵉ siècle*. Paris: Éditions du Seuil, 2008.

Weart, Spencer. *Nuclear Fear: A History of Images*. Cambridge and London: Harvard University Press, 1988.

Žižek, Slavoj. "Passions of the Real, Passions of Semblance." In S. Žižek, *Welcome to the Desert of the Real*, 5–32. London and New York: Verso, 2002.

Internet:

Lemarchand, Frédérik (2011). "De Tchernobyl à Fukushima: une pédagogie du désastre." *Echos*. http://lecercle.lesechos.fr/economie-societe/international/221134016/tchernobyl-a-fukushima-pedagogie-desastre (accessed 2 May 2012).

Manfred Zaumseil and Johana E. Prawitasari-Hadiyono

Living in a Landscape of Risk in Java/Indonesia

1 Introduction

Threat, danger and risk are realities people have to cope with. Mainstream risk research analyzes both how people cope individually and socially and construct the future. However, we encounter problems in their values and belief systems (Zinn 2010). In the past, a deficit perspective dominated, especially in psychological research. The aim was to find out what sort of bias in perception and mental processing was responsible for distorting reality. Meanwhile, the ability of people to judge the relevance of risk has since been re-evaluated. However, in risk management and risk governance, the opposition between real risks and their biased, subjective perception still remains dominant and interconnected with opposing methods such as technical approaches and psychological measuring.[1]

More recently, the understanding of risk has become more complex and complemented by ways of social construction influencing just what counts as objective reality as well as what constitutes the subjective impression. Zinn (2010, 13) for the present discussion notes: "The belief that risks are objective, subjective as well as socially constructed might be seen as common ground in risk research." But the relative importance of the elements in this triangle is a matter of debate.

2 Objective Risk

Objectively assessing and judging risk is based on evaluating the likelihood of dangerous, threatening or harmful events and the vulnerability of those exposed. In these terms, "A risk is the product of vulnerability and endangerment" (Wisner et al. 2004, 51). In the definition of risk in the context of disaster research, a "hazard" as an environmental event is related to the lack or capacity of persons to cope with it within their living conditions. By relating a hazard to

---

1 In biographical research, there are approaches to exploring the construction of certainty at the individual level (Zinn 2004). Less is known about the collective level.

people's scope for coping with it attains the meaning of a danger or a threat. In sociological and psychological risk research, the term "risk" is often applied to the presence of an obvious environmental threat (such as the presence of a nuclear power plant or an active volcano) without explicit reference to the vulnerability of the people exposed. In this case, the risk assessment is reduced to the likelihood of the hazard happening. The objective, measuring approach also gets applied to the assessment of the vulnerability of a given social unit and the environment affected (Wisner et al. 2004). This mainly constitutes a technological approach which proceeds from risk assessment to the management of risk. This nevertheless represented a shift to people's capacity to protect themselves and becoming resilient (see Norris et al. 2008) rather than just analyzing the vulnerability that limits them.

Resilience has been shown to be based not only on "objective" knowledge, which is usually represented by experts claiming scientifically gained and universally valid knowledge. Another source of knowledge in the evaluation of hazard and vulnerability has been valued as experience based local knowledge (see Bankoff 2003; Dekens 2007).

In the Western discourse, there is some dispute about what phenomena should be regarded as risks. Beck (1992) has advanced the idea that the new social uncertainties and the individualization of risk pose a major threat in late modernity. Tulloch and Lupton (2003) argue that these grand claims of theorizing lack empirical support and propose research strategies to investigate risk in everyday life and in the social and cultural context.

There is a tendency towards reification of this objective and universal approach to risk that may obscure the questions of the subjective as well as social, cultural and political processes which endow meaning to risk. Henwood et al. (2010, 3) portray the methodological problems arising "when researchers define research situations from the outset in terms of universal notions of risk" and they recommend valuing diverse ways of producing risk knowledge.

3 Subjective Risk

Subjective risk is how people perceive and interpret as well as act upon a risk. Research has shown that most people exposed to a seismic threat and who are aware of it do not much to reduce their vulnerability (Solberg et al. 2010). From a Western perspective, this may seem counterintuitive. This assessment has been thought to be related to a so called "optimistic bias". "White males" have been found to be prone to this bias and members of so called "collectivistic cultures" with an emphasis on a relational, interdependent self-concept are said to be less optimistic (Palm and Carroll 1998). Harris and Hahn (2011) reexamined the methods of unrealistic optimism research and revealed how the results of studies demonstrating unrealistic optimism could parsimoniously be viewed as statistical artifacts rather than demonstrations of some genuine human bias.

Solberg et al. (2010) in their review about seismic hazard adjustment demonstrate that most of these studies act on incorrect assumptions: the studies assume that the most important causes of seismic adjustment can be conceptualized as mental processes that take place primarily within the individual mind. These processes are further assumed to be primarily consciously available cognitive representations of risks, norms, control and so on. Thus, individualist and rationalist assumptions have dominated the research agenda (1673).

The authors convey how the causal chain does not start rationally from risk perception to produce a risk concern and leads from there to a special practice (seismic adjustment). Instead, the adjustment is a practice related to norms in a social and cultural context. "Issues of cultural identity, power and trust need to be brought centre stage" (1674).

Zinn (2010) points out the different conceptions of the subject in risk research. The individualized risk subject is thought to take cognitive and rational decisions but typically biased through different influences on perceptions and mental processing. Now, this has been replaced by a subject with a softer rationality that works quite successfully with heuristics (see Gigerenzer 2008), intuition and emotion (see Slovic 2010). These psychological models are general theories about human functioning. Zinn (2010) argues for the introduction of a biographical perspective because everyday practice in confrontation with risk is formed by life circumstances. This opens up subjectivity for the biographical, historical and social context and interconnects with the social construction of risk.

4 The Social Construction of Risk

Individual ways of coping and living with risk are socially embedded and framed by the interaction of the social and cultural context along with the physical environment. Within the social and cultural arena (see Long 2001), many different actors influence the discourses about what constitutes a risk and how to deal with it. Media, politics, economics, different social groups and people at risk themselves shape the public and personal representations on the background of cultural heritage. Especially today in the context of a disaster and the often following efforts of risk management, there are many external agents in the arena of humanitarian aid (see Hilhorst and Jansen 2010), national and local aid influencing the emerging discourses about risk. In the following, we present different social constructions of risk and show what different assumptions and interests are interconnected with them.

Bankoff and Hilhorst (2009) illustrate how different actors in the Philippines use different policies of risk for their respective interests. The state policy involves transforming a disaster into an abnormal and singular event. This covers the root causes of the harmful consequences and supports the political goal that deems it sufficient to restore the community to its previous state without changing anything in the social order. Non-governmental organizations, however, adopt a quite different conception of risk. They perceive the disastrous consequences of a hazard as the outcome of a bad policy by the state which needs changing.

Bankoff (2003) and Bankoff and Hilhorst (2009) suggest that the state of normalcy of a secure and ordinary life is a Western myth. Instead, Bankoff (2003) presents data about the Philippines which reveals that disasters are a fact of life and a frequent life experience in the Philippines: Learning to live with hazard and coming to expect disaster is necessarily part of the daily routine in the islands. There is a long history, from at least the seventeenth century, of mutual assistance associations and networks at the local level devoted to sharing risk and spreading the effects of misfortune (Bankoff and Hilhorst 2009, 689).

This understanding may make it necessary to reverse the perspective on risk. The question would be which form of risk adjustment, which psychological, social and economic balance is maintainable in the evident presence of threat and what amount of certainty is necessary for everyday living?

The singularization of disasters enables their exclusion from the normalcy of everyday life. People are potentially overburdened by the constant awareness of the presence of a severe threat and prefer instead to adopt an optimistic outlook towards future. This may take place simultaneously as Bankoff (2003) demonstrates by identifying how preparedness towards natural hazards gets rooted in practices, stances, values and "local knowledge".

This opposition of disasters not only drives their exclusion from everyday living but people may also find risk management and disaster research itself constitutes a special construction of the problem of risk. The biased subject or the biased community has to be elucidated by knowing experts about the probability of the impending natural hazards to prepare for and attempt to control. Implicit in this model is the separating and opposing of humanity and nature, the former taking control over the latter in a particular idea of expertise and a peculiar idea about the singularity of disasters and the certainties of ordinary life.

A historical view (as applied by Bankoff 2003) or an evolutionary perspective as proposed by Wulf (in this volume) uncovers the presence of catastrophes as factors in the man-environment interaction and possibly as a motor of development processes. In the environmental change literature (see Brown and Westaway 2011; Pelling 2010), we find theoretical models claiming to explain under which conditions disasters produce a conservation or a transformation in the social order. Pelling and Manuel-Navarrete (2011) deploy case examples to show the distinction between different forms of resilience supporting rigidity or transformation. In relation to this, they differentiate between a form of risk management that stabilizes the status quo and another championing innovation as "adaptive management".

In the anthropological and historical view on human-environment interaction (Casimir 2008), we find a different perspective on the problem. Casimir points to the early cognitive distinction within Judeo-Christian and later Cartesian thinking and the categories of "humanity" and "nature", which is are rather peculiar within the background of the many different meanings of environment and nature. Taking the example of dealing with volcanic eruptions in Indonesia, Schlehe (2008a) describes in this same volume how nature is constructed in Java. It is not seen as opposed to culture but as parallel to human society and there are close associations between cosmos, morality and social conduct. Hence, a disaster provides a source for the construction of manifold meanings and the interpretation and manipulation of social and political relations. What is interesting in the background to the Bankoff and Hillhorst ideas about the "normalcy" of risk and threat is what Schlehe (2008a, 278) calls the ambivalence towards nature: "people experience powerful awe-inspiring and destructive forces in nature as well as life-giving qualities. In general, the 'dangerous' sites in nature are not perceived in a purely negative way, they are considered spiritually endowed and sacred."

Summarizing, we see that the construction of risk is undertaken by many actors with different interests and forming the politics of risk. The notion of risk is closely connected with a network of related basic assumptions about reality and core stances and feelings about being in the world perceived whether as a safe or as dangerous place. This derives from the construction of nature and cosmos and the range of human agency within it. In the international framework of risk management, the opposition of an objective explainable environment and biased subjectivity serves to educate people how to prepare themselves against possible hazards. There is not enough knowledge about what people in different cultural contexts feel, think and do about their being or not being threatened and how the presence of hazards is part of their everyday social life and their psychological ecology.

Additionally, in most risk research, the religious frameworks, which are of great importance in most parts of the world, are broadly neglected. Another important frame of reference are the narrative and discursive structures which form experience as Toshio Kawai reports in this volume for the Japanese context. These religious and socio-cultural framings are closely connected to psychological dispositions and stances determining how severe events are accepted and dealt with.

In the following sections, we seek to show how the presence of an eventually persisting threat is included in the feeling, everyday living, the practice and future of people in Java, Indonesia, who suffered an earthquake several years ago.

5 Risk from Environmental Hazards in Java, Indonesia

In this contribution, we apply selected data and the results of a larger project,[2] in which we present a model of the whole process of coping with disaster and with disaster aid. For the purposes of this broader research, the disaster which took place in 2006 in the region DIY of Yogyakarta (Bantul) and Central Java (Klaten) in Indonesia provides an example with which to explore the intertwined issues of coping and aid. The earthquake that occurred in the early morning of 27 May resulted in the destruction of 280,000 houses and the deaths of almost 7,000 people. Three villages, which vary in distance from the epicenter of the tremor, have been chosen as research sites. To collect the data, we conducted narrative and guideline-supported interviews with villagers and aid workers, focus-group interviews and field observations during multiple short field trips starting three years after the earthquake. Opportunities for feedback on the village and regional levels provided a different set of data. In 2011, additional data was collected about the future of the villages through a participative research strategy.

---

2 The research project entitled "Individual and collective ways of coping long-term with extreme suffering and external help after natural disasters: meanings and emotions" was funded by Fritz Thyssen Stiftung from November 2008 to October 2011. See Zaumseil et al. 2014.

Grounded in our research we developed the concept of "risk inclusion", which may be seen as a part of coping with a persisting threat and with the future. Risk inclusion takes place in a special physical and social environment, which we call the landscape of risk.

In the following section, we approach the hypothesis that in such a landscape of risk there is the formation of a locally specific way of dealing with the environmental threat, which shows up in everyday practices, in social relations, in values and religious orientations and in psychological dispositions. We term this way of dealing with risk in disaster-prone areas "risk inclusion". We believe this differs in Java when compared for example with Japan or the Philippines and it is subject to change.

We develop the concept of risk inclusion by exploring the meaning given to the earthquake and the ways for individuals and their society to cope with it. Furthermore, we apply results from participative research about the future of the community and their inhabitants to discuss how this interrelates with risk.

6 The Landscape of Risk

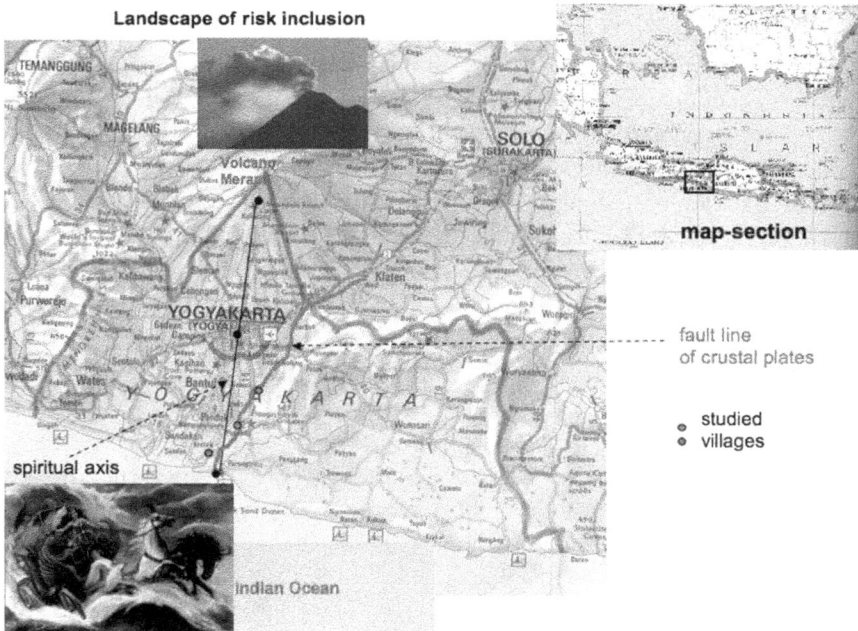

Fig. 1: Yogyakarta

In the local geography of risk, we find an inscription of history and meaning: the province D.I. Yogyakarta is composed of the town Yogyakarta and its surrounding rural hinterland. The town with 500,000 people in the middle part of Java is a vivid center of Javanese culture, arts, tradition, modernity and science. There is a large student community attending its 80 universities and with the dominance of a modern Muslim middle class. The whole Yogyakarta province including its surrounding rural area contains 3.3 million inhabitants with 90% Muslims, some Christians and a Buddhist and Hindu past. The Sultan is the head of the government and is a businessman as well as the head of the traditional court with its ceremonial role.

The palace of Sultan Hamenkubuono X lies in the mid-point between the active volcano Merapi with its Volcano Ghosts to the north and the Indian Ocean, governed by Ratu Kidul, the mystical Queen of the Sea, to the south (see Schlehe 2008b). Beginning with Senopati in the sixteenth century the Sultans are symbolically married to Ratu Kidul and "legitimated and protected by both spirit kingdoms" (279) against the dangers of both the volcano and the sea. This duo of spirits from the mountains and those from the sea receive offerings from Senopati and his successors in the yearly Labuhan sacrificial ceremony. Thus, the Sultan is believed to be responsible for the spiritual balance regarding the forces of the ocean and the volcano (see the spiritual axis in fig. 1). Although modern Muslims refrain from this as superstitious, most Javanese in that area – Muslim and non-Muslim – have this idea integrated into their worldview. In the present period of democratization and decentralization in Indonesia, there is a revival of tradition (Henley and Davidson 2008) in the name of self-determination and the reestablishment of local elites. Schlehe (2010) states that Javanese people are not more "superstitious" than people elsewhere and they know that there are scientific and geological reasons behind an earthquake. They furthermore understand that they are living in a red zone of tectonic activity.[3] However, for many Javanese, causality is also rooted in the spiritual realm (Schlehe 2010).

Within our research project, we interacted with the recent history of social change and the impact of hazards that turned into disasters in the region as well as in Indonesia as a whole. There is a long history of disasters – big volcano eruptions (the biggest in Sumbawa 1815 and the eruption of Krakatoa in 1883) and earthquakes at sea. In recent history, Mt. Merapi near Yogyakarta erupted in 1994 and in Oct. 2010 during our data collection. However, our research project focused on the long term consequences of the earthquake in the southern

---

3 According to the results of Takeshi et al. (2009), the fault may lie slightly further to the east of the Opak fault line in the picture – what seems less to be not so important in this context.

part of D.I. Yogyakarta in May 2006. 17 months earlier, there was the big tsunami hitting the coast of Aceh/Sumatra (2,200 km away from Yogyakarta) in December 2004 with a death toll of 170,000. In July 2006, there was another tsunami once again at Pangandaran, Java and in September 2009 another earthquake in Padang, Sumatra.

7 The Meaning of Earthquake

In our Bantul case, there is a general shared representation understanding earthquakes as warnings or reminders (*peringatan*) or as tests (*cobaan*), examinations (*ujian*). Less often, they are perceived as an expression of the wrath of Allah (murka Allah). The meaning often gets explained in parallel to a scientific understanding (e.g. living in a zone of dislocation of crustal plates). Mostly the idea of reminder, admonition and test does not directly relate to the cause of the disaster or its harmful consequences. Admonition in our data is understood more generally – it concerns the basic relationship between humankind and the almighty or the supernatural.

Although personal meanings attributed to admonition are rather diverse and discovered to be held as deeper wisdom (*hikmah*) (see below), a common feature is the message that humans are basically without control and power in relation to the almighty. Thus, the admonition, warning or test aims at the faith in and commitment to the higher being(s) in the face of divine almightiness. The spiritual lesson from the earthquake aims at surrender rather than (primarily) control. This falls entirely outside the control paradigm of risk appraisal or instrumental coping (Lazarus and Folkman 1984). People see themselves in a sort of dialogue with the supernatural acting through nature. This position within the world and the cosmos includes the possibility humans basically are beings without any everyday understanding of natural hazards and unable to control the natural, supernatural, social/political and personal worlds.

8 Local Key Concepts of Coping and Risk Inclusion

8.1 Personal Coping

Coping is the way in which action relates to the understanding of a threatening situation. To go into more detail about local ways of coping with risk, let us make the acquaintance of Mrs L.:

Fig. 2: Mrs L.

Mrs. L. is married, 31 years of age and has five children. The eldest daughter (at the time of the earthquake, aged 10) died under the debris of the house. Her other daughter (in shorts) showed deviant behavior some months after the event. Mrs. L. had another baby after the earthquake. The family is rather poor compared to its neighbors and owns no rice field.

In the picture, Mrs. L. is sitting three years on from the earthquake in front of her small house (still without plaster.) She says:

> Well, as a matter of fact, I surrender (*pasrah*) like this, whatever it is, I accept it (*nrimo*). No matter what is going to happen, I will accept it. I prepare myself (*siapkan diri*). I should

be prepared – if I suffer a loss or not. If there is another disaster, I may die. So, I will be prepared, as they say, this place is quake-sensitive [...] Maybe it will not be long before it happens again. In fact, people should do something and make an effort (*usaha*), as far as we can, right *Mrs.*? But yeah, although we make the effort, I am destined to have limited fortune [...] We really should accept what happens (*nrimo*) and always acquire self knowledge, true! As for me, well, I've limited fortune. [...] We really should accept what happens in God as being fair. He has given me a bitter test (*cobaan*). But he also gave me a new baby. The deeper meaning and wisdom (*hikmah*) for me is that I just love my children more and, in my opinion, the deeper meaning (*hikmah*) is that I'm a better person than before.

Through Mrs. L, we obtain a basic understanding of the key terms which signify a psycho-spiritual stance: *pasrah* denominates a trustful surrender combined with *nrimo* as a basic acceptance of what happens. It means gaining strength, serenity and inner peace by accepting one's own powerlessness and limited knowledge. It appears as the opposite of gaining immediate environmental control by instrumental behavior. However, this stance fosters pragmatic action and effort (*usaha*) within given limits. Without this trustful surrender and devotion to the almighty, there is no support from this higher power or being (Allah/God). There is a basic trust in the benevolence of the almighty and in a hidden wisdom (*hikmah*) to that happening.

As for Mrs. L. *pasrah/nrimo* is fundamental to being prepared, we find a completely different conception of preparedness compared to international risk management within the Western control paradigm of being prepared in the technical sense. Indeed, those who adhere to the model of effective environmental control may misunderstand *pasrah* and related concepts as fatalism.

Mrs. L. expresses a devotedness to the benevolent almighty power or being and a trust in a hidden wisdom (*hikmah*). From there, she makes a connection with action. This trustful surrender endows the strength and serenity for effort and pragmatic action within given limits (*usaha*). Compared to the dominating Western perspective, there is an over-determination of powerlessness (or limited power) that translates into trust and therefore stands for a general positive outlook. The relationship between trustful acceptance and surrender on the one hand and efforts and problem solving actions (*usaha*) on the other hand is mediated by the general orientations and virtues qualifying the mode of action. Ideally, this involves a stance of inner peace (*tentram*) and tranquility that in turn implies control of negative emotions. The practical, situation-specific action (for instance, helping others, ensuring survival, etc.) is infused with and carried along by the attitude and positive feelings that result from the simultaneous creation of a relationship with the spiritual and the social surroundings. Furthermore, virtues are simultaneously realized through this action, such as *dengan ikhlas* (whole-hearted and willing), *semangat* (with dedication), *sabar* (patient), *santai* (serene) and *menata hati* (control of oneself, self-will).

The limits of pragmatic action are perceived as set by limitations imposed by the lack of resources, power relations within the villages and social rules. These limits refer to the root causes for an unequal exposure to hazards and may only ever be overcome by changing the given social order.

8.2 Social Coping

The above mentioned personal stances and virtues are closely connected to the social world and to collective action. The quality of *tentram* (peaceful, quiet, calm) can be applied to the inner world (*batin*) as well as to one's outer social relationships (as *tentram lahir*). This closely relates to the Javanese idea of *rukun* meaning harmonious integration and the regulation of conflicts by a sort of guided communal harmony (*rukun*). The conjuration of and commitment to *rukun* is a strong force of communal control, which is often functionalized as conserving the given social order. At the same time, this may be understood as a strong bond of communal cohesion when threatened by danger.

The same function holds the tradition of mutual support (*gotong royong*) in Javanese communities. There is a vivid discourse in our data about the function, the obligation, the strength, the possible decline of *gotong royong* about coping with disaster, which demonstrates the importance and salience of the concept for the people in the villages.

Finally, there is a discourse about local wisdom in which all the stances, orientations and virtues described above are constructed and identified as "Javanese".[4] The official report on the post earthquake rehabilitation and reconstruction program for Yogyakarta and Central Java (Tim Teknis Nasional 2007) comes under the heading of "Living in a disaster prone area relying on local wisdom". In this report, local wisdom is seen as a special tradition of social coping. *Gotong royong*, the tradition of mutual help, connects to modern principles of community-based disaster risk management (Davis and Murshed 2006; Davis 2011), empowerment, participation and decentralization in the common discourse. The report states that "local long standing traditions of deliberation for consensus" (Tim Teknis, vi) had been the foundation of a sort of guided need assessment in local groups and decentralized distribution of money. Local wis-

---

4 We discussed the construction of the "Javanese" in another publication (see Zaumseil and Prawitasari-Hadiyono 2012 and Antlöv and Hellman 2005).

dom discourses convey positive images of their own culture and give rise to local pride. Tim Teknis Nasional (38), as well as the local provincial governor, stated that the people of Bantul showed a better way of coping with the quake than those in Aceh when confronted with the tsunami 17 months previously.

Another key concept of social coping we found three years after the quake was that of a general feeling of safety and social embedding (*merasa aman dan nyaman*) after being confronted by the tsunami, it took 17 months to push back a feeling of being threatened. This was not specifically directed towards a specific threat but was a basic feeling of consistency and trust within the neighborhood, community and the religious congregation and related to the stabilizing and calming effect of the daily routine. There is a partial crossover with the concept of social capital.

Hence, some key concepts of social functioning and presently salient discourses about social relationships and values stand out as answers to the threat posed by living with hazards. Taking Bankoff (2007), both emotional and psychological adaptation as well as the manifold forms of community mobilization and mutual self help all develop in societies in which disasters are frequent life experiences.

We consider that beside the threat of hazards, there are other influences which formed communal harmony (*rukun*), mutual self help (*gotong royong*), the whole complex of local wisdom generating local pride and the general feeling of safety. However, these concepts do fit in with the permanent presence of threat and constitute answers that seem at least adaptive. Their salience points to the inclusion of risk into the social realm.

9 Risk Inclusion

Should we combine all these findings, we see that the inclusion of risk can be seen as a part of a general adjustment to risk in disaster-prone areas. This represents a special stance towards threats and towards vulnerabilities.

- This incorporates a shared representation of the "natural" presence of threats, which must not be a conscious awareness of risk but seen as a part of a general adjustment to them.
- This is counterbalanced by a general feeling of safety, social embedding, everyday routine and positive outlook.
- Both – threats and safety – are seen as in the hands of divine or supernatural forces and can be modified to a greater or lesser extent by personal, familial and or community actions.

We would propose that there are different forms of risk inclusion in different contexts. Toshio Kawai shows (in this volume) that the way of inclusion in Japan differs. There is a variety of cultural handling and of politics of risk issues (see Bankoff and Hilhorst 2009) – such may be included or excluded from the official discourse and from everyday practice. As Bankoff and Hilhorst have shown, treating disasters as singular events can be a form of risk – exclusion from the everyday horizon to preserve the present social order. Risk management and risk control strategies are technical measures including a special planning-rationality often introduced by stakeholders from outside into a disaster prone area where they meet with different local concepts and rationalities, which are often implicit in practices and values. Additionally imported concepts and strategies often do not fit in with the conditions in the villages and compete against local configurations of relevance and the shortage of resources in village life.

Thus, pragmatic action is limited by the local social and political order and the economic conditions. This gets informed by the present psycho-spiritual and socio-psychological background of risk adjustment with a cultural/contextual specificity, which is often poorly understood. These conditions meet in a post disaster context with externally introduced guided efforts of risk management and trauma therapy (see fig. 3).

Components of risk adjustment in Bantul/Java

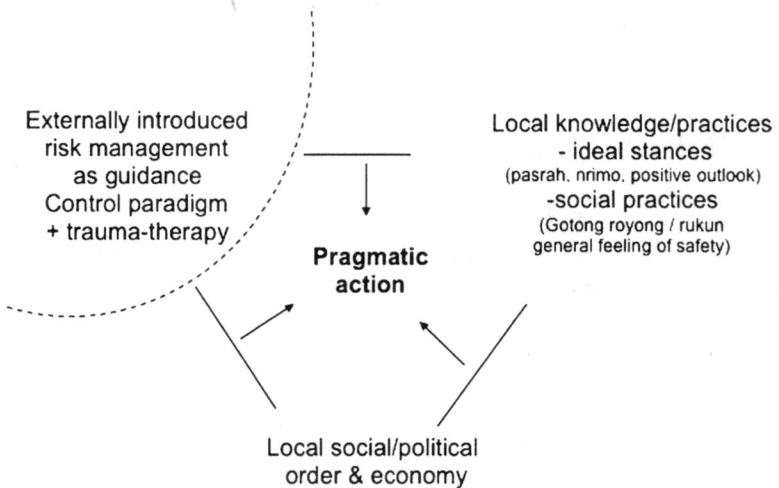

Fig. 3: Risk adjustment in Bantul/Java

According to Solberg et al.,

> Issues of cultural identity, power and trust need to be brought centre stage. Their absence in much of the literature speaks to a profound theoretical limitation, grounded in late modern political and cultural frames. These seek to explain behavior as if it was purely a matter of (ir)rational individual choice in a controllable world (Solberg et al. 2010, 1674).

Years after any disaster, in the area at risk, there is the coexistence and interaction between externally introduced risk management and pragmatic action. The latter is informed by existing local knowledge including ideal stances and social practices. Through national policies, international agents and the media, new ideas about risk control and about trauma as a new concept for enduring personal suffering have been introduced and locally adapted by different actors. The whole process of adjustment takes place within the tension established between conservation and transformation.

After two years of research, we had built up a picture of the perspectives of village people on their coping with disaster around three to four years after the event but we also gained the impression that our coping approach and our quake-related interviewing was a strategy of imposing our questions on the village people. Did the present village life represent long term coping with the disaster or was it rather coping with the hardships of life? How do village people living in a "red zone" of environmental threat (and "knowing" that) deal psychologically, emotionally and practically with the persisting presence of this threat? Or is the relevance of a persistent threat just a construction of observers looking from outside and having a different frame of reference in which risk does not belong to normalcy?

This leads to the questioning of the theoretical research approach and methods. Furthermore, as seen from psychological research, individualized and cognitive access proves rather limited.

Questionnaires do not convey a great deal about everyday living. Interviews give return a better picture but focusing on the threat introduced by the researcher produces a distorted result. Henwood et al. (2010) discuss this question in relation to their research about living near a nuclear power plant in Britain. Referring to Tulloch and Lupton (2003), Henwood et al. (2010) try to place a knowing, reflexive and social risk subject at the center stage of their research. Zinn (2010, 51) comments:

> The approach does not develop a fully biographical argument, which would interpret people's risk perception of nuclear power as part of their more general approach to dealing with biographical risk and uncertainty. For example, instead of examining how people experience their life in general, they focus on "experiences of living close to the nuclear

power station" and thereby frame the interview and the interviewees' awareness in relation to the nuclear power station.

We did not go into the past but tried to overcome our focus on the present threat by catching the view of the villagers about the future through participative research.

10 Turning to Participative Research

To get a grasp of understanding the social and psychological normalization of a situation that outside observers perceive as threatening, we decided to learn from the villagers how they construct, plan and anticipate their future. To avoid the methodological problem of focusing on the researcher introduced context of threat, risk and worrying, we chose a participative research strategy. In the third and last year of our research project, we planned to stimulate pieces of performative social practice about personal and village futures (*masa depan* meaning: period/era ahead) by participative research. Through this, we hoped to gain pictures of the future produced by the villagers themselves (including risk adjustment if made relevant) and less biased by our ideas about what people should do and think about.

The plan also included giving away control about method and content to our local NGO-Partners from ICBC.[5] However, shortly before a planned visit of the German team members to Yogyakarta was to take place and designed to develop a common understanding of the purpose and process of the participatory research with our NGO-partners, the project came to a halt due to the eruption of the nearby volcano Merapi: the spiritual leader Mbah Marijan, who had predicted that there would be no danger for his village, burnt in his house along with his followers. Yogyakarta was covered with ashes and hosted 30,000 refugees from the Merapi area. Our Indonesian team members were involved in rescue and relief activities for the Merapi victims and refugees.

With a three month delay, we were again able to start the participative research. Nevertheless, we were not then sure how the recent disaster in the neighborhood would influence the outlook of the village people towards the future.

---

5 ICBC (Institute for Community Behavioral Change) is a Yogyakarta-based NGO, Director: Edward Theodorus. Cooperating researchers from ICBC were: Nindyah Rengganis, Lucia P. Novianti, Tiara R. Widiastuti.

We addressed existing social groups (youth and children groups, women groups, farmers groups etc.) in the villages to stimulate visions and images about the future. People engaged slowly and soon more and more enthusiastically and actively in producing three films, four booklets, discussions, pictures, interviews, two posters and public demonstrations about the "Masa depan" future. There were presentations of these products at large public village meetings.

All the pictures and activities and products around the future had a positive and solution-oriented outlook and there was no reference to any risk or threat from disaster. The view of the future in the *inter alia* products was:
- a bright time that will arrive,
- something positive to achieve – aspiration (in children/adolescents: what I want to be),
- village development concerns (health, child development, agriculture),
- positive handling of village conflicts within guided harmony (*rukun*).

11 Discussion

We were surprised about the positive outlook and the absence of disaster-related threats, worries and sorrow in the products about the future. The task to produce and present something about the future was understood in the sense of giving a rather general – i.e. non-disaster specific – and encouraging message, a suggestion how to handle community problems in a better way in the future. The future was constructed as a desirable goal. The issue of the threat of disaster or the need for disaster preparedness remained absent. Nevertheless, our qualitative interviews in the previous research period had shown an awareness about living in a disaster-prone area before deliberately having asked for it. However, this aspect of the future was obviously only verbalized when deliberately asked about.

Tekeli-Yesil et al. (2011) reported a rather similar awareness regarding the situation of a predicted earthquake in Istanbul, which they call a "realistic" perception. Therefore, we do think it becomes too easy to impute a denial of threat to the villagers. If we take the perspective of the normalcy of threat proposed by Bankoff (2007) for the Philippines, then uncertainty over living conditions is a sort of normal background to life and some of our interview partners explained to us that there is no point to moving to another place because risk is everywhere in Indonesia and it is better to submit one's own fate into the hands of Allah. The everyday pragmatism, the daily routine with a general feeling of safety, social embedding and a positive outlook seems fostered and not constrained by this (spiritual) background. Hence, risk inclusion via acceptance and trustful surrender while simultaneously striving for the best seems to be adaptive in Java. Additionally, should risk management be introduced, it is important to understand the complex notions of preparedness: worldly orientated pragmatic preparedness might be fostered by spiritual surrender and the acceptance of the immutable. What seems to be an overdetermination of powerlessness combined with the idea of a benevolent higher power creates a contrasting foil of hope and positive outlook and thus supports active engagement in pragmatic action. So, human agency and the conceptions of the inner and outside worlds are constructed in a quite different way compared to Western thinking where powerlessness holds a different meaning.

12 Conclusion

Whenever risk management is imposed by international or national agencies, the constraints of the Western control model should be kept in mind: there is an overdetermination to having the power to control by a top-down logic that often implies a neglect of locally prevalent psychosocial and cultural realities. What may also be overlooked are specific forms of resilience within local forms of risk inclusion. We discussed the distinction made by Pelling and Manuel-Navarrete (2011) between a rigid and a transformative form of resilience. The Javanese case and the form of risk inclusion found this looks like tending towards rigidity and supporting the given social order. However, just as Beatty (1999) and Schwarz (2012) show, there are many hidden varieties and tensions with a transformative potential under the broad roof of commonly conjured general terms and concepts.

Acknowledgment

We thank Silke Schwarz and Mechthild von Vacano for their critical comments to earlier versions of this article.

References

Antlöv, Hans, and Jörgen Hellman, eds. *The Java that Never Was. Academic Theories and Political Practices*. Münster: Lit Verlag, 2005.

Bankoff, Greg. Cultures of Disaster: Society and Natural Hazard in the Philippines. London: Routledge Curzon Press, 2003.

Bankoff, Gregg. "Living with risk; coping with disaster – Hazard as a frequent life experience in the Philippines." *Education about Asia* 12, no. 2 (2007): 26–29.

Bankoff, Gregg, and Dorothea Hilhorst. "The Politics of Risk in the Philippines: Comparing State and NGO Perceptions of Disaster Management." *Disasters* 33, no. 4 (2009): 686–704.

Beatty, Andrew. *Varieties of Javanese Religion: An Anthropological Account*. Cambridge Studies in Social and Cultural Anthropology. Cambridge: Cambridge University Press, 1999.

Beck, Ulrich. Risk Society: Towards a New Modernity. London: Sage, 1992.

Brown, Katrina, and Elizabeth Westaway. "Agency, Capacity, and Resilience to Environmental Change: Lessons from Human Development, Well-Being, and Disasters." *Annual Review of Environment and Resources* 36 (2011): 321–342.

Casimir, Michael J. "The mutual dynamics of cultural and environmental change: an introductory essay." In *Culture and the Changing Environment: Uncertainty, Cognition, and Risk Management in Cross-cultural Perspective*, ed. Michael J. Casimir, 1–58. Oxford and New York: Berghahn, 2008.

Davis, Ian, and Zubair Murshed. *Critical Guidelines – Community-based Disaster Risk Management*. Bangkok: Asia Disaster Preparedness Center, 2006.

Davis, Jan. "Reducing disaster risks 1980–2010: some reflections and speculations." *Environmental Hazards* 10, no. 1 (2011): 80–92.

Gigerenzer, Gerd. *Rationality for Mortals. How People Cope with Uncertainty*. New York: Oxford University Press, 2008.

Harris, Adam J. L., and Ulrike Hahn. "Unrealistic Optimism About Future Life Events: A Cautionary Note." *Psychological Review* 118, no. 1 (2011): 135–154.

Henley, David, and Jamie S. Davidson. "In the name of adat: regional perspectives on reform, tradition, and democracy in Indonesia." *Modern Asian Studies* 42, no. 4 (2008): 815–852.

Hilhorst, Dorothea, and Bram J. Jansen. "Humanitarian Space as Arena: A Perspective on the Everyday Politics of Aid." *Development and Change* 41, no. 6 (2010): 1117–1139.

Lazarus, Richard S., and Susan Folkman. *Stress, Appraisal and Coping*. New York: Springer, 1984.

Long, Norman. *Development Sociology. Actor Perspectives*. London: Routledge, 2001.

Norris, Fran H., Susan P. Stevens, Betty Pfefferbaum, Karen F. Wyche and Rose L. Pfefferbaum. "Community Resilience as a Metaphor, Theory, Set of Capacities, and Strategy for Disaster Readiness." *American Journal of Community Psychology* 41, no. 1–2 (2008): 127–150.

Palm, Risa, and John Carroll. *Illusions of Safety. Culture and Earthquake Hazard Response in California and Japan*. Boulder, CO: Westview Press, 1998.

Pelling, Mark. *Adaptation to Climate Change: From Resilience to Transformation*. London: Routledge, 2010.

Schlehe, Judith. "Cultural Politics of Natural Disasters: Discourses on Volcanic Eruptions in Indonesia." In *Culture and the Changing Environment: Uncertainty, Cognition, and Risk*

Management in Cross-cultural Perspective, ed. Michael J. Casimir, 275–301. Oxford and New York: Berghahn, 2008a.

Schlehe, Judith. "Religion, Natur und die aktuelle Deutung von Naturkatastrophen auf Java." In *Religion und die Modernität von Traditionen in Asien: Neukonfiguration von Götter-, Geister- und Menschenwelten*, eds. Judith Schlehe and Boike Rehbein, 207–234. Berlin: Lit Verlag, 2008b.

Schlehe, Judith. "Anthropology of religion: Disasters and the representations of tradition and modernity." *Religion* 40, no. 2 (2010): 112–120.

Schwarz, Silke. Eine kritische und kultursensible Sicht auf das Mainstreaming von Gender in einem Katastrophenkontext – Umstrittene Gendergerechtigkeitsvorstellungen, ein heuristisches Modell sozialen Wandels und Strategieprozessvorstellungen. *Dissertation* Freie Universität Berlin, 2012.

Slovic, Paul. *The Feeling of Risk*. London: Earthscan, 2010.

Solberg, Christian, Tiziana Rossetto, and Helene Joffe. "The social psychology of seismic hazard adjustment: re-evaluating the international literature." *Natural Hazards and Earth System Sciences* 10, no. 8 (2010): 1663–1677.

Takeshi, Tsuji, Katsuya Yamamoto, Toshifumi Matsuoka, Yasuhiro Yamada, Kyosuke Onishi, Alfian Bahar, Irwan Meilano and Hasanuddin Z. Abidin. "Earthquake fault of the 26 May 2006 Yogyakarta earthquake observed by SAR interferometry." *Earth, Planets, Space* 61, no. 7 (2009): e29–e32.

Tekeli-Yesil, Sidika, Necati Dedeoglu, Charlotte Braun-Fahrlaender and Marcel Tanner. "Earthquake awareness and perception of risk among the residents of Istanbul." *Natural Hazards* 59, no. 1 (2011): 427–446.

Tim Teknis Nasional. *Living in disaster-prone area relying on local wisdom: the post earthquake rehabilitation and reconstruction program for Yogyakarta and Central Java: a guideline for decision makers and program managers*. Jakarta: National Technical Team, 2007.

Tulloch, John, and Deborah Lupton. *Risk and Everyday Life*. London: Sage, 2003.

Wisner, Ben, Piers Blaikie, Terry Cannon and Ian Davis. *At Risk: Natural Hazards, People's Vulnerability and Disasters* (2nd ed.). London: Routledge, 2004.

Zaumseil, Manfred, Silke Schwarz, Mechthild von Vacano, Gavin Brent Sullivan and Johana E. Prawitasari-Hadiyono (eds.). *Cultural Psychology of Coping with Disasters. The case of an earthquake in Java, Indonesia*. New York: Springer, 2014.

Zaumseil, Manfred, and Johana E. Prawitasari-Hadiyono. "Researching coping mechanisms in response to natural disasters: the earthquake in Java, Indonesia (2006)." In *Disaster Politics: Coping Strategies, Representations and Memory*, ed. Ute Luig, 131–172. Berlin: Lang, 2012.

Zinn, Jens O. "Health, risk and uncertainty in the life course: A typology of biographical certainty constructions." *Social Theory & Health* 2, no. 3 (2004): 199–221.

Internet:

Dekens, J. (2007). Local knowledge for disaster preparedness. Kathmandu: International Centre for Integrated Mountain Development (ICIMOD). http://preventionweb.net/go/2693 (accessed 10 May 2013).

Henwood, Karen, Nick Pidgeon, Karen Parkhill and Peter Simmons (2010). "Researching risk: Narrative, biography, subjectivity." *Forum Qualitative Sozialforschung/Forum: Qualitative Social Research,* 11 (1), Art. 20. http://nbn-resolving.de/urn:nbn:de:0114-fqs1001201 (accessed 22 December 2011).

Pelling, Mark, and David Manuel-Navarrete (2011). "From resilience to transformation: the adaptive cycle in two Mexican urban centers." *Ecology and Society* 16 (2): 11. http://www.ecologyand society.org/ (accessed 22 December 2011).

Zinn, Jens O. (2010). Biography, Risk and Uncertainty – Is there Common Ground for Biographical Research and Risk Research? *Forum Qualitative Sozialforschung/Forum: Qualitative Social Research* 11 (1), Art. 15. http://nbn-resolving.de/urn:nbn:de:0114-fqs1001153 (accessed 22 December 2011).

Gabriele Brandstetter
Performance and the Deferral of the Catastrophe Narrative

Naoko Tanaka's Performance-Installation *Die Scheinwerferin* (2011)

In her story "Mein Japan" (My Japan) Yoko Tawada, a Japanese author who has lived in Germany since 1982, writes of the Fukushima disaster as follows:

> When I hear of a disaster, my heart automatically begins to beat more slowly and I become calmer, as though I had taken a sedative. In order to survive a natural catastrophe, one must not get into a panic and start imagining terrible things. It is evident that I have unconsciously learned this attitude in Japan as a survival technique (Tawada 2011).

How are images of a disaster constituted? How are its fore- and after-images narrated and commented on? And what forms of their embodiment and physical-emotional reactions are available to overcome them? Yoko Tawada's description corresponds to the actions, experiences and general bearing of people in Japan after the tsunami of March 2011 – a kind of level-headedness, solidarity and social order in the face of chaos and disaster, which were registered with astonishment in Western sociological commentaries and media discussions. No panic, no resort to states of emergency to prevent plundering,[1] but "discipline and what seemed to be stoic resignation," as the cultural historian Florian Coulmas found (2011, 44).[2] In an essay inquiring into the lessons of the disaster, Paul Carter writes that the civilizational shock triggered by the Japanese tsunami exposed our "lack of narratives, the failure of our prophecies" (Carter 2011, 49). What was needed was a "propaedeutics for the unforeseeable", which meant "incorporating disaster in the here and now" (49). What might such an embodiment consist of? In what way would it bring *other* experiences into play beyond those provided by rational explanations, political discourses or media

---

1 A phenomenon that was observed in the natural catastrophe of Hurricane Katrina. See, on this issue, Gil 2010, 195–206.

2 This question was widely debated by columnists. In an article in the *Frankfurter Allgemeine Zeitung*, Stefan Schulz (2011) conjectures that the calculations of the risk and the (abstract) figures of forecasts on the hazards of a worst case scenario – a "principle of mathemagic" – ultimately caused the "state of affairs in Fukushima [to lead] to a new composure, by making us realize that it is terrible but ineluctable?"

footage? To rephrase the question, is there a poetics, an aesthetics of disaster? And what key images could represent a disaster like the tsunami and that of the nuclear reactor in Fukushima? Does not this disaster and its media presence confirm Paul Virilio's thesis that in the twenty-first century the "original accident" (cf. Virilio 2007) no longer exists and that these "disasters" follow an "invention" of the "technological unconscious" (10): a repetition which – as in the case of Fukushima, whose place in a "logical progression" from Hiroshima and Chernobyl was frequently remarked upon – already has a place in a "Museum of Accidents" (23) for an archaeology of disasters?[3] If it is art that legitimizes the authenticity of the images through the power of imagination – thus creating the possibility of remembrance – what role is played by the 'mediation' and 'narration' of the disaster? For the latter is always a contingent event that bursts in upon a concrete social situation. The imaginative faculty is the basis, the motor of the images and narratives. For the *Writing of the Disaster*, which Maurice Blanchot (1995) makes the subject of his reflections, is a social construct and depends upon contemporary discursive practices. The "writing of the disaster" means both communicating and prophesying disasters and "the writing done by the disasters – by the disaster that ruins books and wrecks language" (ix). "The disaster, unexperienced. It is what escapes the very possibility of experience – it is the limit of writing. This must be repeated: the disaster describes" (7). Such a "de-scribing" is the starting point for art and its capacity to transform the images and rhetoric of a disaster. How can art generate a subsequent, a "restorative fiction" to the "unleashing of violence" (Capeloa Gil 2010, 205)? And in what way do images and embodiments contribute to an experiential space in which narratives carry out a future-directed work on (past) experiences of catastrophic destruction?

The view extends – in a situation in which the narratives and images are published and circulated from an inevitably subsequent perspective – not only into the future (of the twenty-first century), but also backwards – into a history in which different and yet similar narratives of violence and disastrous wars prevailed: as at the beginning of the twentieth century. The religious and cultural historian, Hans-Jürgen Heinrichs speaks of the "disastrous Modern Era". The ruins of the disastrous Modern Era, the nuclear power stations, testify to the relationship between feasibility and sacrifices (cf. Heinrichs 1984, 74).

---

3 See, in reference to this, the article by Capeloa Gil (2010) referring to Virilio.

Ever since Aristotle, the definition of the imagination and the imaginative faculty[4] has included the ability to inwardly conjure up not only pleasurable fantasies but also images of horror. In his major study *Zur Geschichte der Einbildungskraft* (On the History of the Imaginative Faculty), Dietmar Kamper (1981) stresses that the catastrophic element is a key component of an aesthetics of *posthistoire*. Recent research in neurophysiology and brain studies confirms that the imagination mediates between sensuous experience and intelligence, between 'sensory awareness' and mental/cognitive images. Seen in this way, the imaginative faculty is a central factor in 'embodied fantasies'. It structures the relationship between man and his body and also his relations to his stock of images, myths, sagas, and memories of the culture in which he lives. This complex process of a constant transformation between the inner images of a 'self awareness' and an outwardly directed 'attention awareness' develops in paradoxical loops when the hazards of a catastrophic experience loom. At this point, the imagination becomes a protective screen against the traumatizing violence of a naked reality. At the same time, the fantasy images cover over the trauma of fear and horror and take its place. The subsequent *images* of the disaster and its narratives of peril take the place of the disaster. It is precisely this linkage between 'imagination', 'sensuality' and 'embodiment' that gives rise to works of art, putting 'reflective modes of experiencing' and 'critical modes of narrations of the disaster' to the test.

The example that I would like to present a close reading of in this connection could be seen as an allegory of the 'embodied fantasies', as a *visio* of the catastrophic.

I would like to take the performance installation *Die Scheinwerferin* by the German-based Japanese artist Naoko Tanaka[5] as an example of exploring the ways in which images of disaster are embodied and the superimpositions of images and retrospective images that make the subsequent nature of narratives

---

4 Translation note: In German, the term *Einbildungskraft* is part of the tradition of a theory of imagination in philosophy and aesthetics.

5 Naoko Tanaka, *Die Scheinwerferin*; Concept, visual/sound design, Performance: Naoko Tanaka; Dramaturgic collaboration: Mariko Harigai; Duration: 30 min.; Supported by: POLA Art Foundation and Japan Foundation; 14 January 2011 (premiere) and 15 January 2011, TANZTAGE BERLIN/Sophiensäle, in co-production with PACT-Zollverein Essen. Visual Artist Naoko Tanaka studied painting and fine art at the Tokyo Fine Arts University and in 1999 came to the Arts Academy of Düsseldorf in Germany with a scholarship. There, she was co-founder of the artist collective Ludica, with whom she has been creating stage performances and dance installations. In 2010, Ludica was invited to Dance Platform Germany. Cf. Anonymus 2011 (program *TANZTAGE BERLIN 2011*).

of disaster visible. Tanaka's performance was conceived *before* the Fukushima disaster, having premiered in Berlin's Sophiensäle on 14 January 2011. In this installation, the performer shows an imagined journey in distorted shadow images that are projected onto a large split screen. It is a traumatic story with an autobiographical core. We learn from the programme that Tanaka is using images of "her personal story of an eating disorder"[6] which loom out of her memory like a horror trip: 'images' of an individual crisis, indeed a disaster, become – in the context of the national disaster – the earthquake, the tsunami and the threat from the accident in the Fukushima nuclear power plant – a powerful visual evocation of a general threat. In this way Tanaka's performance installation was *subsequently* seen as an ominous warning of the disaster to come. The same thing happened with the choreographed opera *Matsukaze*, composed (after a Noh play by Zeami) by Toshio Hosokawa and choreographed by Sasha Waltz, which was also interpreted in a new and different way after the occurrence of the 'real disaster' in Japan. After 11 March 2011, the dark side of the Noh play was seen in the light of the tsunami.[7] The meaning of deferral – as a difficult question of evaluating decisions, as a re-construction of the 'factual' and as a coping strategy which makes use of the 'fictional' – is clearly exemplified in the following texts: first in Evan Osnos's (2011) critical and admirably researched article in the *New Yorker* on the "fallout" from Fukushima; and secondly in an article by the well-known Japanese director Toshiki Okada (2011) who, looking back on 11 March 2011, wonders what it might mean to "threaten reality with fiction".

What imaginations and fantasies do the images of Tanaka's installation conjure up? And how are they embodied?

It starts with the spectators entering the hall like a room installation. They pass a table at which the performer is seated with her head on her arms as though asleep. Above her hangs a naked light bulb which flickers very brightly, as though something were out of order. Under the table and round about it all sorts of objects lie scattered (fig. 1). They seem to be a chaotic jumble and yet are also like an elaborate and fragile landscape that has been put together out of everyday objects and things such as knives, forks, spoons, discarded toys, the

---

6 http://www.sophiensaele.com/produktionen.php?IDstueck=962 [21.02.2012].

7 When interviewed afterwards, Sasha Waltz said: "Needless to say, in the last days I have pondered whether I, as a consequence of this catastrophe, in regard to 'Matsukaze' as well, should reconsider everything and respond. 'Matsukaze' evokes the heaving seas, the waves; images of the tsunami come to mind" (Waltz et al. 2011, 74). And in the press – also with regard to Fukushima – the end of the opera was interpreted as a "call for reconciliation with Nature" (Herzfeld 2011, 26).

rails of a model railway, lattice work and a tangle of dried-up branches – a bizarre collection of bric-à-brac covered with the dust of forgotten childhood days. Only after all the audience have taken their seats, thus turning the setting of the 'walk-in installation' into a theatrical space of 'stage' and auditorium, does the performance begin. Tanaka rises from the table and directs her gaze to a dummy lying on the table which uncannily resembles her. With a small but very powerful torch – a mini-spotlight – she examines the dummy all over (fig. 2). As in an anatomical experiment, she penetrates the physical image of her double. Thus begins a journey into her own innermost world: by projecting her fantasies, fears and ideas onto her likeness and the surrounding space, she excorporates the 'embodied fantasies' of her traumatic experiences. Tanaka uses the spotlight to explore under the table. The light bobs along, illuminating the objects from various angles. Knives and forks and the other things, the dried twigs and trellis work, are projected on the back wall of the room in the manner of a shadow theatre. They appear as a monstrously prolific dream landscape. The images and sound collage of rattling subway rhythms suggest a dangerous journey, expressed in unforeseeable images, morphings, enlargements and obliterations, flashing by and disappearing (fig. 3). Finally, Tanaka stands at the edge of this installation/formation and illuminates herself with the spotlight from various angles, as she had done at the beginning with her dummy double. Finally, she sticks the torch into her mouth: her head and neck seem illuminated from within in an eerie reddish glow and her physiognomy is distorted as the boundaries between inside and outside have been shifted (fig. 4). The light goes out. The memorial theatre of an 'embodied fantasy', in which the performer is transformed into a "Schein-Werferin" into a disaster story, is over.

Fig. 1

Fig. 2

Fig. 3

Fig. 4

The word used as the title of the performance, "Scheinwerferin", is hard to translate. The term implies several levels of meaning, which play with the semantic field of 'seeing' and 'seeming'. First, in a technical sense, *Scheinwerfer* means 'spotlight', i.e. the powerful lighting equipment used in the theatre [Ger. *Schein* can mean either '(shining) light' or '(illusory) appearance'; Ger. *Werfer* means 'thrower', as in *Speerwerfer* – 'javelin thrower', tr.]. The feminine version of the term [the German ending *-in* corresponds to the English ending '-ess' in *waitress* or '-ette' in *usherette*, tr.] is unusual, a neologism constituting an anthropomorphizing of the device, the object 'spotlight': an 'embodiment' of the thing in female form. A spotlight throws light on other people/things, while normally remaining invisible itself. *Scheinwerferin*, in the *feminine*, embodied form of the term, also suggests that someone here is casting a *Schein* about them while themselves standing in a shining light like an apparition (in a narcissistic way, though also in the aesthetic sense of Adorno's concept of "apparition"). And, finally, *Schein*werferin also refers to the conditions of the presentation in the theatre and the production of (performative) images: *Schein* in the sense of illusion. The *Scheinwerferin* casts light on a *Scheinwelt*, a scene of fantasy.

All these aspects of the title word "Scheinwerferin" are visually activated and embodied in the performance installation. In the words of Phillip Zarrilli: the vision, the *visio* of the *Scheinwerferin* is "enacted" as a physical process (cf. Zarrilli 2007 and 2004).

Tanaka shows images of her physical fantasies and exhibits herself by transforming herself into the 'pointing' (indexical) light (the "Scheinwerferin"). Its question is directed at the *Schein* in its secondary meaning of 'illusion', that transient empty surface concealing a different reality, e.g. the machinery of the theatre, or the inside of embodied feelings. And Tanaka is also exploring the possibility of touching the feelings of others through the *Schein*/'appearance' of her actions as an artist. The process of seeing is itself placed in the (spot-)light here. For the observation, physical sensing and fantasizing of images is shown here in manifold scenes and traces. From the dummy through the dissolution in the chaos of the inner journey to the transillumination of one's own body, we see a performance of Tanaka's 'pictorial self'. It is repeatedly made clear how the imagination of one's own body deviates from the performer's body that we see in action, which implicitly shows the difference between body image and body scheme that is so characteristic of anorexia and eating disorders.[8] Trauma and dangerous self-fantasies transmit and embody themselves in the hallucinatory images of an urban jungle. The imaginative capacity is presented in an

---

8 On body scheme/body image, cf. Krois 2009.

'enacted process'. What seems to be of interest here is the *embodiment* and *presentation* of the inner images.[9] Daily movements – which are learned via cultural body techniques, such as walking or eating with a knife and fork (or with chopsticks) – are seized upon only to have their automatic routine interrupted. The embodiment, the mimetic assimilation of body techniques in a process of 'image-ination' is quoted in the performance. At the same time, however, a break appears in the repetition and transmission – in the sense of that idea of 'embodiment' and 'enactment' that Phillip Zarrilli describes for the body practice of performance as follows:

> When one initiates doing a form of movement or enacting an acting score, one's relationship to each individual repetition of that same form or structure is "similar" yet "different" [...]. As one learns to inhabit a form or a structure of action, one is attuned to an ever-subtler experience of one's relationship to that structure (Zarrilli 2007, 645).

Embodiment is seen by Zarrilli not as a completed action but as a process that is in flux and situated in the in 'between' of what has been experienced and what is to come, between habitat and non-habitat as a space of possibility. Tanaka's performance takes place between such boundaries – and reveals it as a danger zone. For here, the boundaries between *imagines agentes* of a hallucinatory anxiety scenario of the soul, the boundaries between body and environment are blurred. Their blurring and explosive breaching leads to imaginations of the 'inside-to-outside': an ex-corporation of the catastrophic.

Thus, the performance opens up two versions on the subject of 'images of the catastrophe' and 'embodied fantasies'.

In the *embodiment* of *Scheinwerfen* (spotlighting), the process of enacting images becomes self-reflexive. The spotlight, which shines inward and outward, 'throws' images onto the projection screen in a back-reflection – a process leading to a re-vision of the terms 'image' and 'vision'. The range of 'image' (image or picture) goes beyond the optical dimension to include a moving embodiment. This process of 'enactment' incorporates the kinesthetic perception of the spectator. The 'self-awareness' of the beholder of these body-images and phantasmatic image-journeys is affected by transmissions of empathy and "kinesthetic imagination" (cf. Reynolds 2007, 185–211; Foster 2011, 6–14, 126–218). The spectator plunges into the flickering shadow images: into a vision of the chaotic. At

---

9 Within the visual arts and aesthetics, the concept of 'embodiment' is discussed in philosophical terms of aesthetic experience, ecological significance and self-consciousness (cf. Crowther 1993). My argument focuses on embodiment as a mode of kinesthetic, synesthetic and performative transposition.

the same time, s/he observes the 'showing' of what the conditions of this per-formance are: the work with the 'spotlight' – as a 'sine qua non' of the theatre. The projection of the images, their emergence and disappearance; the *artificiali-ty* of the disaster images; their media construction – their illusionary, transient superficiality as well as their physical in-depth dimension.

This finally raises the question of what access this opens to the imaginary. What images of a collective imaginary (as described by Castoriadis 1984)[10] pro-vide a basis for the fantasies and flickering shadows of this autobiographical story of a trauma? Tanaka's installation alternates between an embodied experi-ment on herself and the images of a working through (in the Freudian sense) of her trauma as projected by light media into hypertrophe. In this shadow theatre the rigid, forgotten objects are re-animated: they are set in motion by the "Scheinwerferin" with a 'proto-life', like the 'Undead'. This is the backdrop against which Tanaka's installation could subsequently be read as a disaster scenario: the performance – with its structure of 'doubling' and 'embodiment', of excorporation of traumatic fantasies – offers itself for a "re-thinking of catas-trophes".[11] Not because it offers a reproduction or a representation of *real* disas-ters.[12] It merely reflects those structures which are shaken by disasters: the so-cial behavior patterns for dealing with risk, the gestures of rational and scientific mastery of nature and technology; and finally the culturally conveyed embodiments of behavior in view of a 'hazardous future': panic or composure and acceding to a community of fate, as related in Yoko Tawada's story "Mein

---

10 In his psychohistorical study of the "social institution of the imaginary", Cornelius Casto-riadis (1984) advances the thesis that bourgeois (Western) societies, which see themselves as rational and pledged to the idea of political, technical and economic progress, are based on imaginary structures which subliminally erode this self-image.

11 See the conference thus named in Freiburg, 4–6 May 2011 (FRIAS); see website http://www.frias.uni-freiburg.de/institute/interdisciplinary-encounters-en/catastrophes [4 November 2011].

12 Exactly one year after the Fukushima disaster – as a reminder of the event and as a warning against a 'hazardous future' – a whole series of performances of remembrance were held. Particular attention was attracted by artistic reflections on urgent open questions. Works like the string quartet "Limited Approximations" (premiere Lucerne September 2011) by the Austri-an composer Georg Friedrich Haas, the theatrical works of the Japanese director Toshiki Okada ("The Sonic Life of a Giant Tortoise"), or Daisuke Miura ("Castle of Dreams") made clear that a rational discourse in academia, politics and the public media is not enough to express those aspects of the catastrophic which lie beyond the boundaries of the imaginable and utterable. See in this connection the programs of the playhouses in Munich and Berlin in March 2012: Spielart-Festival Munich: http://www.spielart.org/programm/, the event entitled "No Go Area Japan – A Year After Fukushima" at Berlin's Deutsches Theater: http://www.deutsches thea-ter.de/home/sperrzone_japan/ as well as the works of the artists Nina Fischer and Maroan el Sani: http://fischerelsani.net/ [all accessed 6 March 2012].

Japan" (2011), written after the Fukushima disaster? The double meaning of "Scheinwerferin" shows us that the 'embodied fantasies' can be both true and deceptive.

References

Anonymous. *TANZTAGE BERLIN 2011. 05. BIS 15.01* (program). Berlin: Sophiensäle, 2011.
Blanchot, Maurice. *The Writing of the Disaster*. Lincoln, Nebraska and London: University of Nebraska Press, 1995.
Carter, Paul. "Freuds Salto. Turbulenz als Mechanismus des Eros oder die Lehren der Katastrophe." *Lettre International* 93 (2011): 48–52.
Castoriadis, Cornelius. *Gesellschaft als imaginäre Institution. Entwurf einer politischen Philosophie*. Frankfurt am Main: Suhrkamp, 1984.
Coulmas, Florian. "Freiheit und Sitte. Aspekte der Naturkatastrophe in Japan." *Lettre International* 93 (2011): 43–46.
Crowther, Paul. *Art and Embodiment. From Aesthetics to Self-Consciousness.*" Oxford and others: Clarendon Press, 1993.
Foster, Susan Leigh. *Choreographing Empathy: Kinesthesia in Performance*. London and New York: Routledge, 2011.
Gil, Isabel Capeloa. "Die Herren der Bilder. Die Verbildlichung der Katastrophe bei Walter Benjamin und Spike Lee." In *Rahmenwechsel Kulturwissenschaften*, eds. Peter Hanenberg, Isabel Capeloa Gil, Filomena Viana Guarda and Fernando Clara, 195–206. Würzburg: Königshausen & Neumann, 2010.
Heinrichs, Hans-Jürgen. *Die katastrophale Moderne. Endzeitstimmung, Aussteigen, Ethnologie, Alltagsmagie*. Frankfurt am Main and Paris: Qumran Verlag, 1984.
Herzfeld, Isabel. "Du heißt Wind, und du heißt Regen. Ein Nô-Spiel, getanzt: Sasha Waltz inszeniert Toshio Hosokawas 'Matsukaze' im Schillertheater." *Der Tagesspiegel*, 21 037 (17.07.2011): 26.
Kamper, Dietmar. *Zur Geschichte der Einbildungskraft*. Munich and Vienna: Hanser, 1981.
Krois, John Michael. "Beginnings of Depiction. Iconic Form and the Body Schema." In *Individualität und Selbstbestimmung*, eds. Jan-Christoph Heilinger, Colin G. King and Hector Wittwer, 361–376. Berlin: Akademie Verlag, 2009.
Okada, Toshiki. "Die Wirklichkeit durch Fiktion bedrohen. Der Regisseur Toshiki Okada beschreibt, wie der 11. März 2011 sein Theaterverständnis auf den Kopf stellte." *Theater der Zeit*, no. 10 (2011): 18–20.
Osnos, Evan. "The Fallout. Seven months later: Japan's nuclear predicament." *The New Yorker* 32 (2011): 46–61.
Reynolds, Dee. *Rhythmic Subjects. Uses of Energy in the Dances of Mary Wigman, Martha Graham and Merce Cunningham*. Alton and Hampshire: Dance Books, 2007.
Schulz, Stefan. "Zahlen zum Schönrechnen. Warum wir aus Katastrophen nichts lernen." *Frankfurter Allgemeine Zeitung* 185 (11.08. 2011): 27.
Tawada, Yoko. "Mein Japan." *Christ und Welt* 12, 17.03.2011: 1.
Virilio, Paul. *The Original Accident*. Cambridge and Malden, MA: Polity Press, 2007.

Waltz, Sasha, et al. "Beauty in Impermanence. Sasha Waltz and Toshio Hosokawa in conversation with Ilka Seifert regarding the choreographic opera 'Matsukaze', creative processes and the catastrophe in Japan." *Matsukaze* (program), 73–87. Berlin: self published, 2011.

Zarrilli, Phillip. "Toward a Phenomenological Model of the Actor's Embodied Modes of Experience." *Theatre Journal* 56, no. 4 (2004): 653–666.

Zarrilli, Phillip. "An Enactive Approach to Understanding Acting." *Theatre Journal* 59, no. 4 (2007): 635–647.

Claudia Benthien
The Subjectification of Disaster in Video Art: *Incidence of Catastrophe* by Gary Hill

1 Introduction: Gary Hill and Maurice Blanchot

Gary Hill's video art work *Incidence of Catastrophe* treats an extremely personal, mental as well as physical, experience: the subjective, somewhat unmotivated and in the end (seemingly) lethal suffering of an individual. Consequently, the present article will not deal with catastrophe and disaster on a collective level, on the level of communities. However, the artist chose the term 'catastrophe' to figure in the title of his work and in the piece itself one finds both frequent allusions to natural disasters and an imagology of calamity that is to be found in 'real' catastrophes.

Hill's monumental videotape work of 43:51 minutes duration was produced in 1987–88 in cooperation with a British TV-channel but never actually broadcasted (cf. Sarrazin 2000a, 220). It is one of the last single-channel works of the artist before he turned to multiple-screen video installations. It is also the first videotape piece for which Hill, who usually performs himself in his works, was not behind the video camera but rather engaged a professional cameraman for the recording (219). *Incidence of Catastrophe* is an in-depth and personal examination of the French writer, literary critic and philosopher Maurice Blanchot's first novel *Thomas the Obscure* (*Thomas l'obscur*, 1941) – a highly self-reflexive and experimental piece of literature. Hill's video is not to be considered as an audio-visual "translation" but rather as an autonomous "*transcreation*" of the novel (Machado 2000, 165), following the artist's main intention of "transform[ing] literary and philosophical themes into immediate sensory experience" (Duncan 1995, 69).

Blanchot's novel amalgamates previous literary trends of the twentieth century. Both the aesthetic innovations of authors such as Marcel Proust, Franz Kafka and James Joyce on the one hand, and the literary concepts of the Surrealists, the Existentialists and the *noveau roman*, on the other, are integrated and worked through in a peculiar way (cf. Theisen 2009, 623). In his text, Blanchot tries to adapt their psychological content, formal and narrative experiments as well as their philosophical ideas and, above all, to surpass the writing techniques and modes of perception of the Surrealist authors. Consequently, the book does not consist of a coherent plot and is loosely presented as a dream narrative of the title character. This dream's depiction is distinguished by frequent inconsistencies and gaps. A stream-of-consciousness technique is systematically applied in that the distinctions between dream and reality, thinking and perceiving, presence and past become blurred (cf. p. 624).

Thomas, the title figure, is reading the book in which he himself appears as protagonist. Not only is he split into two entities – for instance, he sees himself swimming in the ocean (cf. Blanchot 1973, 9) –, he is also mirrored by a female figure with whom he has a romantic relationship and who suffers a mysterious death. This young woman, Anne, is the second main character of the novel and builds the narrative center of five of the altogether twelve chapters (cf. chs. VI–X, 41–86). Hill, however, eliminates the female counterpart and focuses solely on the story of the male figure, Thomas, whom he depicts as a restless individual being driven by his own "inconsolable solitude" (Blanchot 1995, 10).[1] In this way, Hill is in line with an age-old Western tradition that considers the single white male as the archetype and paradigmatic figure of reflection for philosophical questions on such fundamental concepts as 'being', 'death', 'nothingness', 'sense', 'appearance' and the like. For Blanchot, it is similarly the male figure of Thomas who suffers a terrible illness being caused by reading a book alone at night in his room:

> The illness is in one of the senses: Thomas suffers from a flood of meaning, he is inundated by experiences. Each physical and sensory effect, from the sensation of being in water, to the language he hears at the dinner table on to the act of reading, overwhelms him. Meaning is disintegrated as it is being constructed; the mind, like the body, is apt to drown, sink in the initial pages, until it breaks down completely [...]. While there are no explicit references to the book Thomas is reading in Blanchot's text [...], in Hill's tape, the artist as Thomas is reading Blanchot's book and uses the language in it as "triggers" for representation, and for the progressive loss of control over body functions (Sarrazin 2000b, 79–80).

The video camera moves over some of the book pages as if it were not only reading them but also actually touching and feeling them physically. It focuses on single words, phrases and pages in such detail that one recognizes the protagonist to be without a doubt reading the 1973 translation of *Thomas l'obscur* by Robert Lamberton (cf. Blanchot 1973).

In his novel, Blanchot works with *mise en abyme* structures that often appear in such highly auto-referential and anti-mimetic literary texts as the *nou-*

---

1 At one point in the videotape, at the prominent dinner party scene, a nameless young blond woman appears briefly, standing opposite the protagonist and then sits down at the table simultaneously with him (23:26–25:00). This woman could be an allusion to Blanchot's female protagonist but this is not clearly marked and the figure plays only a minor role in the narrative.

veau roman and post-modern narratives (cf. Wolf 2008, 502–503). With the term *mise en abyme*, one describes a (literary) recursiveness and self-referentiality in which at least one element (of the content or of formal nature) appears analogously on a subordinate level (502–503). In *Thomas l'obscur* this element is the act of reading – *la lecture* – being one of the principal themes of Blanchot's corpus of writing (cf. Hill 2002). Hill's video, in which the artist himself performs as a *Wiedergänger* (*revenant*) of the literary figure Thomas, amplifies the *mise en abyme* structure exposed by Blanchot: the protagonist here is situated *vis-à-vis* an increasingly physical text, in which he is about to drown.[2] Contrary to the permanent reading of Hill's character, the literary protagonist Thomas is only reading in one – albeit highly significant – scene, during which he has the feeling that the words are becoming alive and are in fact observing him. This corresponds to Hill's own experience with Blanchot's book, when he claims that while reading he had "the rather strange experience of the book reading me – we were reading each other somehow" (Hill quoted in Lestocart 2000, 233).

> He was reading with unsurpassable meticulousness and attention. [...] The words, coming forth from the book which was taking on the power of life and death, exercised a gentle and peaceful attraction over the glance which played over them. Each of them, like a half-closed eye, admitted the excessively keen glance which in other circumstances it would not have tolerated. [...] The pleasure in fact became very great. It became so great, so pitiless that he bore it with a sort of terror, and in the intolerable moment when he had stood forward without receiving from his interlocutor any sign of complicity, he perceived all the strangeness there was in being observed by a word as if by a living being, and not simply by one word, but by all the words that were in that word [...]. Rather than withdraw from a text whose defenses were so strong, he pitted all his strength in the will to seize it, obstinately refusing to withdraw his glance and still thinking himself a profound reader, even when the words were already taking hold of him and beginning to read him. [...] For hours he remained motionless, with, from time to time, the word "eyes" in place of his eyes: he was inert, captivated and unveiled. And even later when, having abandoned himself and, contemplating his book, he recognized himself with disgust in the form of the text he was reading [...] (Blanchot 1973, 25–26).

This central passage of the novel is visually represented in the videotape and 'read' both by the (subjective) camera and the protagonist, the latter being depicted as desperately longing for an understanding of this highly self-reflexive

---

2 Another *mise en abyme* is the audiovisuality of the videotape piece itself since it is explicitly thematized and reflected through its own means and by a rivalry – or rather "deconstruction" (cf. Sarrazin 2000b, 71) – of sound and image and their relationship. "One of the themes in the work of G.H. consists in grasping the mechanisms of this phenomenon of splitting, duplication, self-reference in language according to various modalities by connecting them through vision, and in apprehending the meaning of language insofar as it questions itself" (Lageira 2000, 38).

meta-text. Hill's protagonist, contrary to the Thomas figure in Blanchot, is essentially and continuously engulfed by his book, but also its very materiality and physicality, which is being explored by the camera and his hands in great detail. Blanchot's novel surrounds him and as soon as he opens his eyes from sleep he starts reading.

Gary Hill is one of those early video artists who often use their own body as material in their artworks. However, in the piece we are dealing with here, the fact that he himself is performing as the main character fulfills a much deeper intention: "The choice for me to be in it had more to do with my own experience of reading *Thomas the Obscure*. I didn't want to verbalize it; I wanted to 'be' this. I wanted to put myself against this text" (Hill quoted in Sarrazin 2000a, 218). It is of significant importance that the protagonist does not articulate himself verbally in the video: Hill's Thomas is presented as a passive and mute recipient of literature, of nature and of society. Whereas Hill uses his own voice and poetic language in many of his other works – for instance in *Primarily Speaking*, *Around & About* and *Mediations* – it is a decisive choice in *Incidence of Catastrophe* not to have the main figure articulating himself verbally at all. He is, on the contrary, forced into an incessant, mute conversation with his book and with nature. Referring to a long-standing tradition established in the Middle Ages and gaining prominence during Early Modern book-printing, silent reading is related to interiority and the mind (cf. Benthien 2006, 37–42) as well as to fantasy and 'madness'. The ostentative receptivity and passivity of Hill's protagonist may be contextualized within Blanchot's theory of disaster, which is in fact largely a theory of passivity and endurance. He writes, for instance:

> We might coin a word for the absolute passiveness of total abjection – *le subissement*, which is [...] simply a variation of *subitement* [*suddenly*], or the same word crushed; we might invent that term, le *subissement*, in an attempt to name the inert immobility of certain states said to be psychotic, the *patior* in passion, servile obedience, the nocturnal receptivity of mystics – dispossession, that is, the self wrested from itself, the detachment whereby one is detached from detachment, or again the fall (neither chosen or accepted) outside the self. Still, these situations, even if some are at the limit of the knowable and designate a hidden face of humanity, speak to us hardly at all of what we seek to understand by letting this characterless word be pronounced, *passivity* (Blanchot 1995, 15).

Hill 'transcreates' such complex and paradoxical concepts of passivity and endurance into the aesthetics of his video art piece on Blanchot's novel. The above quote is taken from a second book by Blanchot that is important for this video: *The Writing of the Disaster* (*L'écriture du désastre*, 1980; first English translation 1986). This essayistic work consists of theoretical reflections, written in a fragmentary and aphoristic style, on the interrelation of writing, suffering and death, with continuous direct or indirect reflections on the Shoah (cf. Blanchot 1995, 6, 28, 47, 81–84).

The Writing of the Disaster repeatedly challenges the notion that we can fully represent history, or catastrophe, as comprehensible and complete. For Blanchot, historical testimony is by its very nature flawed, troubled and incomplete. That is not to say it is false, or that it is not important to us. But rather, that we cannot assume to understand the full scale of the disaster, the catastrophe, or the traumatic reality of historical events. The horror of a given event can never be made present again, neither by means of witnesses, nor texts, nor photographs: it remains obscene (off-stage) and beyond our grasp (Tranter 2009).

Blanchot links his reflections on disaster to the act of writing and basically claims disaster to be something essentially "unknown" (Blanchot 1995, 5), "unexperienced" (7) – as that "what escapes the very possibility of experience" (7), it cannot be "de-scribed" (6) and therefore constitutes "the limit of writing" (7). In claiming that "[t]he disaster does not put me into question, but annuls the question, makes it disappear – as if along with the question, 'I' too disappeared in the disaster which never appears" (28), Blanchot refers to the double problematics of the inability to experience and to represent extreme suffering, which has been emphasized in recent trauma theory. The response to 'trauma', as the overwhelming experience of a sudden, catastrophic or violent event or accident that cannot be fully grasped as it occurs, can take shape in the often delayed, uncontrolled appearance of intrusive phenomena such as flashbacks, nightmares, hallucinations, or other repetitive symptoms (cf. Caruth 1996, 11, 91). The traumatized person is haunted or possessed by the traumatizing event and cannot find closure; the event itself is ever-present and subject to a continuous reenactment (cf. Caruth 1995, 4–5; 1996, 92). It can neither be completed – lived through – nor fully be repressed and is therefore suffered time and time again, so that "immediacy, paradoxically enough, may take the form of belatedness" (Cauth 1995, 6). It is precisely this inversion and questioning of linear temporal structures – "since the disaster always takes place after having taken place, there cannot possibly be any experience of it" (Blanchot 1995, 28) – that Blanchot refers to and that Hill adopts in his video's aesthetics and narrative.[3]

---

3 Hill reflects on the possibilities of temporal structures as follows: "Many of my early single-channel video pieces were in a sense 'system-performances' that generated their *own* time in relation to *real* time. There are really so many folds in time involving media, feedback, delay, writing, speaking, and the body. Time becomes more like a Möbius band or Klein bottle without an absolutely 'real' side" (Hill quoted in Quasha and Stein 2000, 244).

2 Hill's Aesthetics and Narration of Disaster: Exemplary Scene Analyses

Thomas's experience of the sea and of swimming is one of the leitmotifs of Blanchot's novel and this "water imagery" (Sarrazin 2000b, 80) is translated aesthetically into Hill's video. In the opening sequence, images of immense waters are to be seen, overflowing the land, breaking the sand banks – strong and violent ocean tides and waves that fill out the complete screen (fig. 1; cf. 0:00–2:12[4]). The soundtrack is likewise intense, producing an audiovisual impression of losing ground, of being taken away by the flood, especially in connection with images of the text being flooded with water (fig. 2). In the second section, one sees and hears monumental book pages being turned (fig. 3). Extreme close-ups of pages are crosscut with images of ocean waves that flood both the readable text and the material texture of the book. Hill uses several forms of visual editing: fades to black, dissolving, frame edits. But the most interesting operation in this regard is the use of soft-edge wipes to produce images of waves rolling over certain lines of a page that is 'read' simultaneously by the camera and the protagonist alike (cf. Sarrazin 2000b, 80). The acoustic dimension – for example, an intense rushing and roaring of the sea, supplemented by the sound of the heavy and hectic turning of book pages, or a loud industrial, urban or even warlike sound – dominates the ominous atmosphere. Scripture and textuality, on the one hand, and nature as something that is violent and impermeable, on the other, penetrate the realm of experience of the protagonist, who is – corresponding to the audience – increasingly incapable of differentiating between dreams, hallucinations and reality. Both the book and the sea are perceivable on the visual and the acoustic 'channels' alike and are similarly contrasted in their atmosphere and intensity. The first images of the book, though, are presented in silence and surrounded by darkness, implying an 'interiority' (cf. 2:22–3:20). However, shortly after the turning of pages itself becomes a very physical and loud action, rhythmized by sound as well as by a back-and-forth of the camera as if the words were coming and going like ocean waves (cf. 4:20–5:10).

During the video's 43 minutes, Thomas becomes more and more tormented by the text, not by what it says, but by what it represents *physically* and by its

---

4 Figures 1–12 video stills from Gary Hill: *Incidence of Catastrophe*, 1987–88. References to the videotape work are provided with temporal codings of the passage 'quoted'. Gary Hill's *Incidence of Catastrophe* can be seen online on YouTube and Video; video art collections that hold a copy are, among others, the Zentrum für Kunst und Medientechnologie (Karlsruhe), the Inter Media Arts Institute (Düsseldorf), and Electronic Arts Intermix (New York).

material threat: the text [...] drive[s] him into a world of nightmares, it becomes a forest of verbal signs where the character gets lost or drowns, a forest that cuts him, penetrates his body so violently that he becomes incapable of controlling himself (Machado 2000, 165).

Fig. 1

Fig. 2

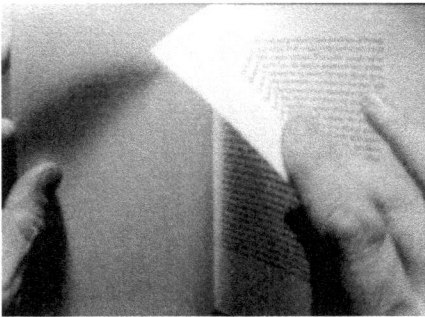

Fig. 3

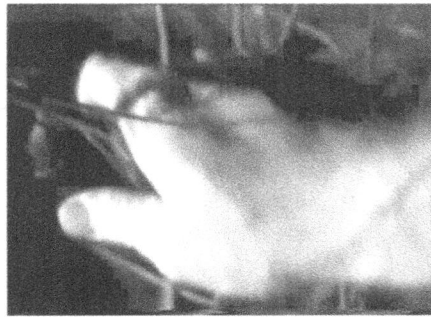

Fig. 4

Only after 'sea' and 'book' are introduced as central 'figures' does the protagonist appear on the scene. He is introduced as the reader of the book and as someone trying to find his way through a dense undergrowth in the dark of the night (fig. 4) – a long scene that stands allegorically for the impenetrability of Blanchot's text (fig. 5; cf. 7:48–12:55). Throughout the video, visions, hallucinations and prevailing darkness hinder the continuous reading and the figure's capacity to understand, to reach 'enlightenment'. He is repeatedly depicted as dead or seemingly dead: for instance, at the end of one sequence showing him reading he lies outside an old cottage, his body caught by the camera eye for the first time through a dusty window full of spider webs (fig. 6; cf. 12:55–13:30).

The (apparently) dead body is then contrasted with the same body sleeping, lying in bed, dreaming; when he wakes up he immediately starts to read the book, only to fall asleep again and vice versa (cf. 13:30–14:49). During this passage we hear strong 'sounds of civilization' from the off: traffic and construction noise, later something that sounds like an airplane starting several times. These sounds permeate the intimacy and quiet of the sleeping and reading situation and insist on the interdependency of interior and exterior, of the mental and the 'real' world.

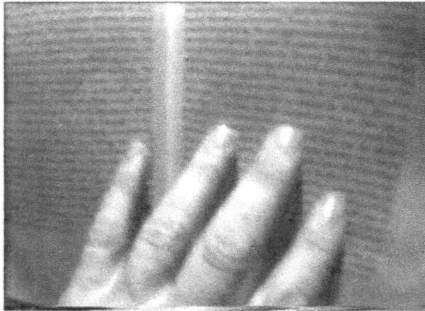

Fig. 5

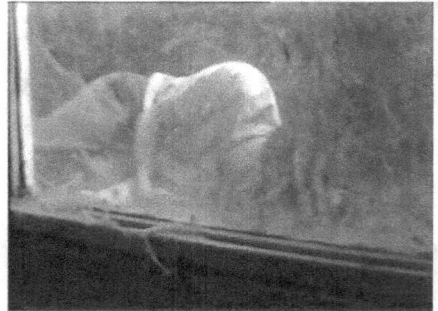

Fig. 6

Fig. 7

Fig. 8

In the beginning of another sequence, the protagonist is again engulfed by his reading of the book (fig. 7; cf. 18:33–19:41), suggesting that the following scene may be something he reads or imagines: a group of seven people having a festive dinner around a table, discussing a non-identifiable subject (fig. 8; cf. 19:41–22:40). In contrast to the passivity and non-articulation of the protagonist, the people at the dinner party talk and interact without interruption. In the narrative mode of 'zero-focalization' the floating camera rotates, slightly above ground

level, around the group that is sitting in a non-definable interior space, surrounded by darkness. Their voices have a certain reverberation, as if under
water, which adds a slight uncanniness or dream quality to the scene. The representation of the dinner party emphasizes two elements closely related by their
orality: eating and drinking on the one hand, verbal articulation on the other.[5]
Suddenly, the conversation stops and a female off-voice is to be heard, articulating the very clear audible question "Have you swum today?" followed by a collective confusion and an embarrassing minute-long silence. This sentence is an
indirect quote from the novel, where it appears twice (Blanchot 1973, 19, 79).

Later on, the dinner party collapses in a literal sense. The main figure (Hill)
continues his solitary autistic reading of the book in the midst of the group of
people dining but then falls backwards in his chair, taking the tablecloth and all
of the dishes with him (fig. 9; cf. 27:35–28:04), a scene repeated later, where all
of the food and tableware sink together with the man into the depths of the
ocean (fig. 10; cf. 31:36–31:58). Another part of this sequence shows the party
itself 'sinking' into madness and surrealism, regressing into a 'pre-civilized
state' (fig. 11). People lose their clothes, they crawl on the floor, play with food,
utter senseless sounds or syllables (cf. 29:34–31:36), until finally the protagonist
remains alone in a crosscut sequence between the bodily states of swimming
and sinking on the one hand and sleeping restlessly on the other, both actions
represented as being surrounded by plain darkness (cf. 32:40–33:34).

The closing passage of the videotape piece contains a certain climax – or rather anti-climax – of this loosely woven narrative. The protagonist gets sick,
fantasizes and surreal images take a hold of him, he nervously observes his
estranged face in the mirror, then vomits and seems to sink again into the ocean
(cf. 34:40–39:12) until his body is to be seen, lying in the sand and low tides. In
the closing sequence, the protagonist is huddled in a fetal position on his bath-
room floor in his own excrement, regressing into the state of an inarticulate
baby, babbling incomprehensible sounds, while a long, bizarre stick, fixed to
the camera, pokes at different parts of his body as if it were a dead animal. This
stick could symbolize the general attempt of the camera not just to observe but
also to practically touch the objects of the material world. In the very final scene, the book pages turn into colossal walls, closing around the naked figure like
a prison cell (fig. 12; cf. 39:12–42:40) and the text of Blanchot's novel "passes

---

5 This theme of orality is highlighted by the next scene to follow, where Hill inserts short clips
of the man reading, of the man searching for something in the book and of his extremely enlarged, open mouth being treated by dental examination tools, while the ocean, strong tides
etc. visually appear within it (cf. 24:22–27:35).

over the walls of the room he is in, as if it had literally taken possession of the body and was dragging it down with it in its disaster" (Lageira 2000, 38). "If the lying man opened his eyes, he would see a mirror that is a text" (Belting 1994, 47; trans. C.B.) – a text that consists of nothing but references to itself. And if video art is – due to its dominant mode of self-portraiture – to be considered, as Rosalind Krauss has prominently argued, as an "Aesthetics of Narcissism" (Krauss 1994), the artist-protagonist here is mirrored in his pre-existing narrative, which not only dominates him but ultimately hinders his general existence (cf. Belting 1994, 47). The "disintegration of speech" (Cooke 1992, 18) results not only in the dissolution of the printed text through water on the visual level but correspondingly in glossolalia on the acoustic level. Ironically, as one learns from Hills *catalogue raisonnée*, the words babbled by the protagonist do acquire meaning if the tape is turned backward: they stand for different body parts (cf. Broeker 2002, 130) – perhaps precisely those parts that are simultaneously being touched by the 'camera stick'.

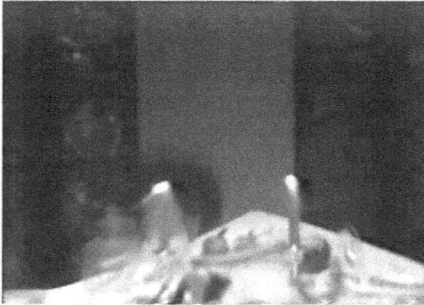

Fig. 9

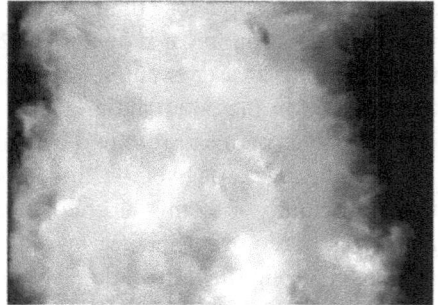

Fig. 10

Fig. 11

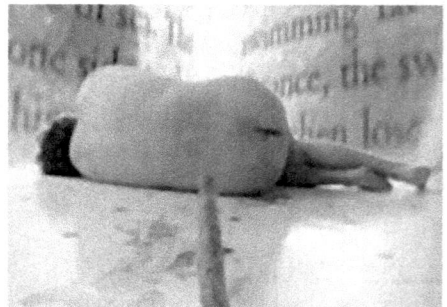

Fig. 12

3 The Subjectification of Disaster – Theoretical Backgrounds and Contextualization

After this close reading of the central scenes of the videotape work, one might ask: on what level (or levels) then, is the theme of disaster treated in *Incidence of Catastrophe* and what should the reference to this category imply?

The ancient Greek term *katastrophé* does not correspond neatly to our understanding of this concept. In classical literature, catastrophic events were usually not classified as such; on the contrary, they were not addressed by the generic term or superordinate concept *katastrophé* (cf. Hebbeker 1998, 13) but by their proper names, as earthquake, flood, fire, dry period, famine, epidemic etc. (cf. Meier 2007, 44[6]). At times, they were also called *pathémata* – for instance by the historian Thukydides – as "distressing injuries suffered by human beings" (cf. p. 51). The composite word *katastrophé (καταστροφή)* is translated as follows: *katá (κατά)* meaning 'down', 'downward' and *stréphein (στρέφειν)* meaning 'turning', 'turning into decline', 'descent'. In general, the Greek term stands for a dynamics of 'turning around', 'reversal', 'inversion' or 'overturning' but also for abrupt social changes such as an 'overthrow' or 'upheaval' as well as for irreversible incidences such as a 'downfall', 'ruin', 'destruction' or 'end' (cf. p. 47–48). Furthermore, in the context of the poetics of drama, *katastrophé* implies a 'decisive turn', especially the inexorable fatal end to a tragedy (cf. Hebbeker 1998, 12; Meier 2007, 48–49[7]), also including the non-tragic resolution of conflicts. Contrary to our notion of catastrophe (or disaster), it was not exclusively used as a term for sudden and singular incidences but also contained the idea of processual occurrence, for instance in analytical drama with its successive uncovering of facts that took place before the dramatic action itself (cf. Immer and Schweikle 2007).

In the modern understanding, however, a catastrophe is a natural or historical disaster characterized both by its suddenness and its destructiveness. The historian Egon Flaig defines it analytically as fitting "extreme events, which firstly do damage to the semantic order in a way that the victims are unable to

---

6 Meier refers, amongst others, to the example of Thykydides and claims that his typology of catastrophes is commonly called *pathémata* list (Meier 2007, 50).

7 Aristotle does not explicitly use the term *katastrophé* but instead *metábasis* for a slow transformation with regard to simple actions, *metabolé* for the sudden reversal that is accompanied by powerful incidences and *peripéteia* for the fundamental turning around of the whole plot on the climax of entanglement; cf. Hebbeker 1998, 12. In his *Poetics*, Aristotle applies the term πάθος ('pathos', 'suffering') in the sense of *katastrophé* (cf. Aristotle 1995, 66f, ch. 11).

cope with that event by a narrative, and which secondly interrupt the continuity of social processes" (Flaig 2007, 35). In modern societies, both natural disasters and manmade catastrophes largely carry the same semantics and rhetoric. Both cause major suffering and are considered to be irreversible and, in their effects and symbolization, singular traumatic events. The latter notion of uniqueness is broached as an issue in the introduction to this volume under the heading of a "critique of singularity", challenging "the narrative behind national images of disaster, i.e. that they remain quite exceptional. Selectivity, repetition, citation, transference, excess, remediatization and self-reflexivity are structural forms constructing a rhetorical framework for the acting out of disaster on the representational level" (Gil and Wulf in the introduction to this volume). One can discuss and criticize Hill's videotape work within this context in its attempt to describe disaster as a highly personal and singular occurrence. It has also been rightly remarked that in modern societies and mass media the term 'catastrophe' – corresponding to other heavy-laden ancient concepts, such as 'destiny' or 'tragedy' (cf. Felski 2008, 3) – is not only used for collective traumata alone but also for individual and private suffering. One has to ask then which dimensions and understanding of catastrophe Hill is implying in his work's prominent title. Since a 'catastrophe' is – at least in the modern notion of the word – considered as an 'incident', as something that suddenly occurs (and only once), turning the world inside out and upside down, Hill's choice of the word *'Incidence' of Catastrophe* is especially revealing. It implies a frequency and repetitiveness of catastrophe that stands in harsh contrast to the modern notion of this term. Within the long video piece there is no singular incident to be identified as the catalyst for the disaster. Corresponding to Blanchot's novel, it is rather a continuous state of suffering that is depicted and embodied by the artist, a slow mental and physical destruction, which is latent at the beginning and gets worse and worse as the work proceeds. Hill thus seems to question the notion of catastrophe or disaster as a singular 'event' in the first place and does so, on a formal level, in the mere length of the piece and, on a referential level, in the title.

Another significant feature of the video is the fact that Hill narrates the story of a specific individual (played by himself), but does so with continuous references to archetypical images and symbols. The crisis and collapse of the protagonist is contrasted with the evocation of a natural catastrophe: that of a tremendous flood. This reference seems to be not so much related to contemporary, 'real' catalysts of disaster, such as global warming and climate change, but to 'timeless' and transcultural images. Even though one may read some latent cultural criticism in the fact that Hill uses industrial or war-like sound in one sequence, it is not explicitly related to the theme of flood, water and drowning.

The artist is, on the contrary, narrating a 'natural catastrophe' that is taking place only for one singular individual – or maybe, on an even further reduced level, in this person's imagination.

It is therefore simultaneously a process of appropriation and subjectification of catastrophe: appropriation in that it is displayed and lived through by a singular person who suffers an enigmatic 'disaster from within' that is only projected onto the exterior world; subjectification in the sense that these destructive experiences seem – in a rather paradoxical manner – to be constitutive for the psyche of Hill's protagonist. This corresponds to the post-structuralist notion, according to which subjectification is considered as something preceding the subject in the same way that individuation precedes the individual. Because subjects are created by way of power-relations they are unable to consciously control or even perceive their subjectivity as such. Also, there are no "autonomous, counter-hegemonic subject positions" (Heller 1996, 92) in the first place. This notion, therefore, broaches the issue of power-relations that are beyond the individual's control – and in this regard perhaps corresponds to the idea of catastrophe. Hill's work therefore does not depict an 'active' appropriation of disaster but something that happens unwillingly, that is passively endured. The Blanchot novel read throughout the video then stands emblematically for the nameless power that dominates and overwhelms the individual with its imperatives.

Gary Hill has remarked that, aside from Blanchot, a second author was important as a background for *Incidence of Catastrophe*: the evolutionist Gregory Bateson (2000) with his monumental work *Steps to an Ecology of Mind* (1972), in which processes of consciousness are viewed as a complex ecological system.[8] In Bateson's view, the world is a series of systems, amongst them individual systems, societal systems and ecosystems and each of them depends upon feedback loops to control balance. All three systems are part of one supreme cybernetic system controlling everything beyond the self or the individual.

---

8 "[I]t has very much to do with *Thomas the Obscure* but also another Blanchot work, *The Writing of the Disaster*. On the other hand, and perhaps less obvious, it had to do with [Gregory] Bateson's *Steps to an Ecology of Mind* in which the processes of consciousness are viewed as an ecological system – putting that very simply, some trees are strong and survive while others die much like ideas. In the work there are several references to this both literal and metaphorical. By the end, and looking back at what the 'protagonist' has gone through I can't imagine any other word but catastrophe! I even thought a little about catastrophe theory and if the point or fold where consciousness breaks could be seen somehow ... and yet consciousness is still consciousness broken or not or so it would seem" (Gary Hill in an e-mail to the author, 30 October 2011). For the significance of Bateson in Hill's videos, also cf. Sarrazin 2000b, 75–77.

Bateson refers to it as the (collective) "Mind". He regards consciousness as the bridge between the cybernetic networks of individual, society and ecology and considers that a mismatch between these systems due to improper understanding will result in the degradation of the entire supreme cybernetic system ("Mind"). Taking these complicated and somewhat esoteric thoughts seriously, Hill's video piece is presenting exactly these three interdependent systems as out of line with each other: the individual is alienated from society and he experiences ecology – the natural world – as both impenetrable and terribly overwhelming.

The videotape then seems a rather bleak diagnosis of the mental state of a human being who is fundamentally out of balance and taken by "the irresistible movement toward an imminent catastrophe" (Blanchot 1973, 51). One may wonder if the irony and playfulness that is so prominent in many of Hill's single-channel pieces and video installations is entirely absent here. Does he really display such an apodictic and mono-linear reading of catastrophe, inspired by Blanchot? Is "literature – and, by extension, language – becom[ing] aggressive and intrusive, leading the reader into a neurotic self-consciousness that is ultimately destructive" (Duncan 1995, 71)? Or is the video artist not, at the same time, questioning this depressive statement through his specific aesthetic means, such as the use of stereotypes, exaggeration, repetition, his choice of an archaic natural imagology and, above all, the use of psychoanalytic symbols (e.g. the notion of regressing to the 'anal phase')?

The protagonist of *Incidence of Catastrophe* can be understood as a personification of catastrophe with all the paradoxes implied: as someone who both gains and loses his subjectivity through the enigmatic disaster that he suffers. Hill himself remarks with regard to this as well as other works of his in which Blanchot's ideas about the suffering or despair created by writing are reflected: "In all these works, language, meaning, and in a sense the vessel from which these flow – the body – are going through an entropic process toward either some prelinguistic state or a process of death – it's rather ambiguous" (Hill quoted in Lestocart 2000, 234). This collapse of meaning – or "breakdown of representation" (Sarrazin 2000b, 75) – that the protagonist in *Incidence of Catastrophe* endures, corresponds to the loss of his capacity to fully experience (or express) the catastrophe as such – or, in Blanchot's own, aforementioned words: "since the disaster always takes place after having taken place, there cannot possibly be any experience of it" (Blanchot 1995, 28). The unusual length of this video work with its continuous repetitions and variations of the same motifs can be understood as an audiovisual examination of traumatic structures that allow neither completion nor repression. Hill's video exposes cyclic and inverted temporal chronology on different levels, most prominently in the turning backward of language in the final scene. Throughout the video work, the inability to understand what is happening in combination with a general fragility of the protagonist's body and psyche leads to the conclusion that the artist does not consider a neat distinction between natural and manmade catastrophes as being of any relevance. Rather, he is much more interested in the complex processes of subjectification of disaster between acts of appropriation and submission.[9]

References

Aristotle. *Poetics*. Trans. Stephen Halliwell. Cambridge and London: Harvard University Press, 1995.

Bateson, Gregory. *Steps to an Ecology of Mind*. With a new foreword by Mary Catherine Bateson. Chicago and London: University of Chicago Press, 2000.

Belting, Hans. "Gary Hill und das Alphabet der Bilder." In *Gary Hill. Arbeit am Video*, ed. Theodora Vischer, 43–71. Ostfildern: Cantz, 1994.

Benthien, Claudia. *Barockes Schweigen. Rhetorik und Performativität des Sprachlosen im 17. Jahrhundert*. Munich: Fink, 2006.

Blanchot, Maurice. *Thomas the Obscure*. Trans. Robert Lamberton. New York: David Lewis, 1973.

Blanchot, Maurice. *The Writing of the Disaster – L'écriture du désastre*. Trans. Ann Smock. Lincoln and London: University of Nebraska Press, 1995.

Broeker, Holger, ed. *Gary Hill. Selected Words. Catalogue raisonné*. Katalog Kunstmuseum Wolfsburg. Cologne: DuMont, 2002.

Caruth, Cathy. "Introduction." In *Trauma: Explorations in Memory*, ed. Cathy Caruth, 3–12. Baltimore: Johns Hopkins University Press, 1995.

Caruth, Cathy. *Unclaimed Experience: Trauma, Narrative, and History*. Baltimore: Johns Hopkins University Press, 1996.

Cooke, Lynne. "Gary Hill: 'Who am I but a Figure of Speech?' " *Parkett* 34 (1992): 17–19.

Duncan, Michael. "In Plato's Electronic Cave." *Art in America* 69 (1995): 68–73.

Felski, Rita. "Introduction. Defining Tragedy". In *Rethinking Tragedy*, ed. Rita Felski, 1–28. Baltimore: Johns Hopkins University Press, 2008.

Flaig, Egon."Eine Katastrophe definieren. Versuch einer Skizze." *Historical Social Research* 32, no. 3 (2007): 35–43.

Hebbeker, Gerhard. "Die Sprachlosigkeit der Katastrophen und die begrifflichen Fassungen ihrer Bedeutung." In *Naturkatastrophen in der antiken Welt. Stuttgarter Kolloquium zur Historischen Geographie des Altertums*, eds. Eckhard Olshausen and Holger Sonnabend, 9–14. Stuttgart: Steiner, 1998.

---

9 The author would like to thank the Hamburg University student research assistants Ole Hinz, Anna-Lisa Menck, Markus Redlich and Lydia White for their help with materials and the manuscript itself.

Heller, Kevin Jon. "Subjectification and Resistance in Foucault." *SubStance* 25, no. 1 (1996): 78–110.

Immer, Nikolas and Irmgard Schweikle. "Katastrophe." In *Metzler Lexikon Literatur*, ed. Dieter Burdorf, Christoph Fasbender, and Burghard Moennighoff, 377. Stuttgart and Weimar: Metzler, 2007.

Krauss, Rosalind. "Video. The Aesthetics of Narcissism." *October* 1 (1994): 51–64.

Lageira, Jacinto. "The Image of the World on the Body of the Text." In *Gary Hill*, ed. Robert C. Morgan, 27–55. Baltimore: Johns Hopkins University Press, 2000.

Lestocart, Louis-Jose. "Surfing the Medium." In *Gary Hill*, ed. Robert C. Morgan, 232–239. Baltimore: Johns Hopkins University Press, 2000.

Machado, Arlindo. "Why Do Language and Meaning Get in a Muddle?" In *Gary Hill*, ed. Robert C. Morgan, 158–167. Baltimore: Johns Hopkins University Press, 2000.

Meier, Mischa. "Zur Terminologie der (Natur-)Katastrophe in der griechischen Historiographie – einige einleitende Anmerkungen." *Historical Social Research* 32, no. 3 (2007): 44–56.

Quasha, George and Charles Stein. " 'Liminal Performance': Gary Hill in Dialogue." In *Gary Hill*, ed. Robert C. Morgan, 243–268. Baltimore: Johns Hopkins University Press, 2000.

Sarrazin, Stephen. "A Discussion with Gary Hill." In *Gary Hill*, ed. Robert C. Morgan, 206–223. Baltimore: Johns Hopkins University Press, 2000a.

Sarrazin, Stephen. "Surfing the Medium." In *Gary Hill*, ed. Robert C. Morgan, 62–90. Baltimore: Johns Hopkins University Press, 2000b.

Theisen, Josef. "Maurice Blanchot: 'Thomas l'obscur'." In *Kindlers Literatur Lexikon*. Vol. 3, 3rd and newly worked over edition, ed. Heinz Ludwig Arnold, 623–624. Stuttgart and Weimar: Metzler, 2009.

Wolf, Werner. "Mise en abyme." In *Metzler Lexikon Literatur- und Kulturtheorie. Ansätze – Personen – Grundbegriffe*, ed. Ansgar Nünning, 502–503. Stuttgart and Weimar: Metzler, 2008.

Internet:

Hill, Leslie (2002). "Entretien: Sur un désastre obscure." *Espace Maurice Blanchot*. http://www.blanchot.fr/fr/index.php?option=content&task=view&id=57&Itemid=40 (accessed 19 August 2012).

Tranter, Rhys (2009). "Maurice Blanchot, The Writing of the Disaster. A Reflection on the French Writer's Seminal Work." *A Piece of Monologue*. http://www.apieceofmonologue.com/2009/11/maurice-blanchot-writing-of-disaster.html (accessed 19 August 2012).

Picture credits:

Figures 1, 3, 4, 5, 6, 9, 10, 11: own video stills, made from a copy borrowed from the Zentrum für Kunst und Medientechnologie (ZKM), Karlsruhe, in 2011.

Figures 2, 7, 8, 12: video stills scanned from the following catalogue: Broeker, Holger (ed.). *Gary Hill. Selected Words. Catalogue raisonné. Kat. Kunstmuseum Wolfsburg*. Cologne: DuMont 2002, p. 130.

Eduardo Cintra Torres
Identifying a Genre: Televized Tragedy

Disasters and catastrophes are recurrent. And yet, they always remain unexpected and leave lasting impacts despite their otherwise ephemeral characteristics. The same happens to their television coverage as they consequently become a rather coarse object of analysis and identification. This article is based upon two very different events that took place in 2001: on 4 March, the collapse of the Hintze Ribeiro Bridge, between Castelo de Paiva and Entre-os-Rios in Portugal, over the Douro river; and on 11 September the terrorist attacks in the United States. For a study on audience reception, a third event was added: the pro-Indonesian militia attacks in East Timor following the 1999 referendum that attained independence. We are not here comparing or evaluating these events that took place on three different continents but rather the televised broadcasts that covered them, one in the United States and the other two in Portugal, involving the full and complete study of the Castelo de Paiva bridge coverage.

Retrospectively, the origins of this genre may be traced to the 1963 assassination of the United States President, John F. Kennedy. This was referred to by Edgar Morin as a "teletragedy" (1964, 81), a coarse object of analysis and events falling within this genre would for example include the Azores earthquake (1980), the explosion of the *Challenger* space shuttle (1986), the San Francisco earthquake (1989), the deaths of King Baudouin of Belgium (1993) and Princess Diana (1997) or the Al-Qaeda attack in Madrid (2004).

These real world events are characterized by society as *tragic* and presented by television in accordance with structures and characteristics that render the successive coverage tragic in genre whether as text or as spectacle. The broadcasts concern the events most intensely mobilizing society extensively and thus displaying the characteristic of what Marcel Mauss termed "total social facts", those social facts that "shake the entirety of society and its institutions" and are simultaneously "juridical, economic, religious and even aesthetic" phenomena (2002, 13).

This necessarily requires a multidisciplinary approach for analyzing televised content intrinsically bound up with a totally tragic fact thus involving not only the study of television *per se*, but "reconstructing a set within which the internal coherence of the society observed emerges" (13). To analyze television, an object of study that appears "colossal, chaotic, complex" in the words of John Hartley, a multidisciplinary approach limits the "injustice" potentially stemming from a single line of research (1999, 19). After all, this is the core of Television Studies, "a youthful field" (Brunsdon 2000, 625) but already established in Anglo-Saxon countries as a branch of cultural studies while advancing only more slowly in Portugal. Competing to contribute to this new field are approaches based on concepts and theoretical frameworks drawn from other areas, such as literary and film studies, social psychology, philosophy, psychoanalysis, ethnography, anthropology, sociology, history and economics. Collectively, they amount to approaches constituting a "body of recognisable and legitimate research on television as a cultural phenomenon" (Casey et al. 2002, vii).

This research project identifies the televised tragedy in accordance with three facets: textual analysis, sociological analysis of the "total social fact" and its impact on audiences in addition to a lesser focus on the institutional and technical frameworks for television broadcasts and approaching the study object through a multidisciplinary approach making recourse to literary, cultural, journalistic and sociological studies, among others.

The fall of the bridge and 11 September are events impacting on the entire community and are experienced as such. Taking up the concepts of Ferdinand Tönnies and Max Weber, we may state that in the case of catastrophes experienced as national, in the case of televised tragedies, *community* superimposes *society* (Tönnies 1979) and affections overwhelm social action inspired "on a *compensation* between interests and rational motives" (Weber 1983, 23). These events shake the very foundations of institutions and instill the concept of crisis. The common and journalistic discourse immediately labels these events as tragedies, nurturing a correlation between the tragedy as a text and theatrical performance and the tragedy as a real world disaster. Hence, this correlation stems from reality being organized, explained and accepted in accordance with a millennial model of literature and performance.

The fact of society transforming into a community that feels itself in crisis provides the foundations for comparing events of such widely different proportions and natures as the bridge collapse and 11 September as what matters is the structure of the event when narrated and shown. This enables the genealogy of the televised tragedy genre to be set out and traced back to the assassination of John F. Kennedy and forwarded to the Madrid 11 March attacks and beyond. In this type of catastrophic event, television takes on a central role in crystallizing the tragic model. Television throws us into the centre of events given its omnipresence in homes, work places and social environments due both to its audiovisual qualities and to it being the leading means of distributing information covering events as well as to the way it enables the viewer to question the relationships of power in a moment of crisis. Given the growing technical means and the competition between broadcasters, television participates in/of the event. Because of its presence, the vision and interpretation of the event by the viewer changes but, furthermore, the dynamics of events subsequent to the original founding tragic fact are primarily established in function of television.

In identifying the televised tragedy genre, we would highlight the following characteristics put forward in an earlier study (Torres 2006):

- The accessibility of technical means, competition between operators as well as the ease of reaching large audiences and in fact generating factually provable broadcast audiences far greater than normal.
- Intensive recourse to live broadcasting and becoming the core of television output interrupting and replacing normal broadcasting.
- Recourse to archetypes, symbols and myths recurrent to the tragic world such as, and for example, the transformation of events into tragic dramas, the classification of events into pre-existing fictions (violence, flooding) and the dramatization of participant characters.
- The recourse to configurations about *destiny* and the *divine* within the framework of the happening, experienced in events and by viewers or imagined by them.
- The open assumption of the tragic nature of events and the human condition inherent to them.
- Features from tragic texts such as units of action, time and place.
- The death and final destiny of the bodies, as well as their eventual display, as a central question in the development of events, with television encountering problems and solutions similar to those faced by the authors of classical tragedies.
- Shift in television journalism style in times of catastrophes, adopting strategies similar to classical tragedies, including emotional discourses without which instead of televised tragedy there would only be the spectacular facet of television that would run counter to the ethics of many viewers given the event relates to the national community.
- The transformation of participants into characters similar in type to those found in classical tragedies. As happens in the latter, we also find that the roles assumed by personalities result from the ongoing actions of the televised tragedy. In the cases studied, the spotlight was thrown on individual heroes (council mayors), collective heroes (the rescuers and the victims, portrayed heroically), family members as spokespersons for the victims and motors of action, witnesses as narrators piecing together events without the intermediation of journalists, the bearers of oracles, scapegoats and guilty parties (real or imagined), political and religious figures, those in suffering and all the essential tragic choir figures (persons present in the locations, onlookers, witnesses, survivors, family members, colleagues, friends and neighbors of victims, all representing the city but without ever substituting

its legality) with the choir serving as a mirror to the audience. Also appearing as characters are the journalists themselves in addition to the television station as a collective entity.

– The televised tragedy genre is also displayed in the need felt by participants, the political authorities, television managers and viewers to provide *closure* to the tragedy, which everybody strives to contribute towards and in conjunction with the intervention of religious institutions and the new, dominant secular institutions such as entertainment and sport. In some tragedies, providing this closure reveals a fundamental concern of power to establish a post-crisis sense of balance.

– Contributing powerfully to the characterization of these transmissions as televised tragedies is their *emotional dimension*. Emotions have taken on greater priority in both journalism and television and, as in other fields of research, have steadily lost their taboo status. In the case of televised tragedy, analysis of the total social fact has to involve the emotional dimension as such represents an integral facet of the study object. There is manipulation and a living of emotions undertaken by participants (broadcasters, characters, receptors) and there is, as has been proven, a direct and unbreakable relationship between reason and emotion that leads the viewer into passing judgment on events about which both factual and emotional information is being conveyed.

In our approach to the emotive charge broadcasts contain, the emotiveness of television viewers and of televised tragedies in particular, this research project sought to portray the various dimensions to televised phenomena: the level of emotiveness experienced by the audience; the mix of information with greater or lesser emotional impact; the relationship between the cognitive and emotive dimensions in the reception of televised messages; similarities between theatrical and televised experiences, the level between the stage and the audience; audience categories by level of emotion experienced; the social conditioners of emotive types, as well as the political dimension to emotions; the influence of broadcasts loaded with a highly emotional charge for the collective; the socialized and the majority accepted emotions, lived and reflected in television output, the approximation between the emotions present in televised tragedy and those present in tragedy-performances, taking Aristotle as our reference as regards fictional narratives and narrating reality (Aristotle, undated and 1991).

In order to understand how such phenomena interrelate with televised tragedies, we carried out a convenience sample survey of 1,329 persons divided up into three sub-samples and freely choosing between responding to the television coverage of East Timor, the Bridge or September 11; stating the extent to which they experienced a range of 17 emotions and also responding to behavioral questions and issuing opinions on the tragic footage seen.

Factor analysis returned an important similarity in the emotional dimension to each of the three independent sub-samples. The emotion of horror always emerges bound up with fear and rage, and with pity and sadness in two of the three events. Joy, happiness and indifference appear associated in the three sub-samples, the same happening with solidarity, sharing and interest. Finally, shame, guilt and disdain tend to be linked together.

Pity and horror, identified as the objective of tragic texts and representation ever since Aristotle, return one of the strongest statistical correlations and the outright strongest in the case of the Timor sub-sample and form the core cluster of the most experienced emotions. The survey demonstrated the existence of a highly similar emotional pattern across the three independent sub-samples.

The five most experienced emotions are exactly the same across the three events: pity, horror, interest, sadness and solidarity, except in the September 11 sub-sample, in which solidarity is replaced by surprise. Despite the differences in the 'content' of the three events, viewers display similar emotional behaviors. These similarities are defining and justify the foundation of a *common category:* in terms of reception, televised tragedy provokes a display of the same emotions and stands out for the non-existence of the same emotions. There is a broad *communion* of emotions and a similarly broad *identification* at the emotional level: televised tragedy unites persons through their emotions. However, this *common category* gains still further consistency on proving that the behaviors and opinions of viewers of each of the tele-tragedies is highly similar.

The three socio-demographic survey variables, gender, age and schooling, do not return any significantly expressive differences within each sub-sample able to provide an explanatory basis for the emotiveness and there is no uniform pattern between the three events when comparing the results of the three aforementioned variables. This fact may confirm that in periods of "collective effervescence" (Durkheim 1968), factors such as social class and educational level and even age and gender shed their importance with the determinant factors driving behavior and opinion stemming from factors upstream such as language, nationality and, even further upstream, emotivity. Hence, we attempted to categorize tele-tragedies audiences not by their variables but according to their respective levels of emotivity in accordance with the *K-means clustering* methodology thereby grouping them by the maximum level of internal affinity based upon the level of expressed emotivity. This process resulted in four clusters, which we defined as highest emotivity, average emotivity, low emotivity and finally a group of low and differentiated emotivity that initially puzzled us before we identified it with Alexithymia behavior, defined by emotional confusion, particularly when confronted by traumatic events, as is our case here.

This categorization into clusters revealed a clear relationship between the level of the emotional reaction of the televised tragedy viewer and his/her behavior and opinion. The same level and type of emotivity broadly corresponds to the same opinion and the same behavior. This means the socio-demographic variables contribute towards differentiating between respondents in normal daily life situations but this differentiation ends up being almost entirely swept away when confronting the extreme situations contained in televised tragedies able to generate what we term *tele-crowds*.

Televised tragedies comply with the principles set down over 160 year ago by the Portuguese novelist and playwright Almeida Garrett on presenting his tragedy *Frei Luís de Sousa* (*Friar Luís de Sousa*): in a "democratic century", he wrote, "readers and audiences [...] want a stronger substance, less spiced and more substantial: they are the people and want the truth. [...] Contemporary drama and novels provide a mirror in which to admire oneself and one's times, the society that is above, below and at one's own level – and the people are destined to applaud as they understand: it is necessary to understand to be able to appreciate and like" (Garrett 1904, 773).

Televised tragedies supply tragic spectacles to the public, with action, truth and emotion, are understandable, constructed piecework into a fictional-like narrative and bringing about immediate live fruition to the sheer weight of numbers in democratic societies. We would highlight: televised tragedy, as with classical tragedy, is a genre belonging to democratic systems and is generally repressed under authoritarian regimes (flooding in the Lisbon region, 1967; the sinking of the Russian submarine *Kursk*, 2000; the Chechen attack on a Moscow theatre, 2002).[1] The consequences for the authorities of the existence of televised tragedy are under debate in open societies and a central concern to what Tamar Liebes identifies as the televised tragedy genre under the name of "disaster marathons" (Liebes 1998).

---

1 We would put forward the hypothesis that in democratic regimes, televised tragedy might be shunned by the televised media in harmony with the political powers. Observed only at a distance, through British media channels, the 2005 terrorist attacks on the London transport system would suggest self-limitation being imposed by television based journalists at least in terms of emotionalizing and in the blanket coverage of the event, and returning to *'infobusiness as usual'* just as soon as was feasible. There are indications that the government and media institutions would have been responding according to some concerted strategy designed for any eventual terrorist attack.

As Goethe had first forecast two centuries earlier, the press, with its real world news would draw on the impact of tragic plots in theatre plays (Goethe 1999, 30–36). What does this mean today when faced by a constant and uninterrupted flow of news in the daily life of a post-sacred world? The spectacle of literary tragedy becomes less necessary to a society saturated with the tragic events of the real world and has been replaced whether by other means of performing the tragic spectacular, or imbued with tragedy as is the case with televised tragedies. What society once internally incorporated through the artistic transcendence of tragic text transfigured into the *common text* – hence, current and live, vernacular, colloquial, informal and common – of live television broadcasting real scenes of real tragedy. Instead of living the "hypothetical suffering" of the mythological dramas, people now gain live access to "real suffering", as if in some Roman arena but nevertheless repressing the pleasure that might perhaps be taken from viewing this "suffering at a distance". Catharsis has been transferred from the performance venue to the living room.

What does it matter – not from an aesthetic or literary point of view, but from a sociological perspective – that "when a hypothesis unfurls in reality there is a factual corruption of tragedy"? (Nuttall 1996, 77). The coverage of tragic events resorts to tragic archetypes and to the same emotional palette deployed by tragedies; making recourse to individual and collective concerns similar to the tragedy; and deploying discursive, structural and spectacular strategies recognizable from the genre of tragedy. Consequently, televised tragedies enable society to accept the unacceptable and make the (in)verisimilitude, integrating into social experience a rupture of values and institutions while nevertheless, through closure, ensuring a return to the equilibrium pre-existing the respective event.

While, in its core origins, the tragedy resulted from a ritual related to the sacred, today, televised tragedy, its corrupted heir, is also a ritual, now post-sacred, but responding to a need to affirm social and community cohesion. As in the tragedy, "the true material" of televised tragedy may also be the social thinking specific to human society (Vernant and Vidal-Naquet 2001, 15) and the reason that both are characteristic and inherent to democratic society.

Both tragedy and televised tragedy relate their own times while doing so from different angles: the tragedy transfigures current themes into fiction while televised tragedy leaves the performance venue behind to make real catastrophes the stage for the new tragedy.

Hence, television has created the new tragedy, the televised tragedy that turns tragic real world events into *reality tragedy,* an elusive genre able to impact on the lives of citizens and audiences.

Therefore, in usurping a dramatic genre, television once again *attains ful-fillment* in its destiny both in historical-sociological terms and in its techniques of immediacy, live, utterly current and "reality", one of the few words that Nabokov finds "as meaning nothing without quotation marks" (1995, 312).

References

Aristotle. *Poética*. 6th ed. Lisbon, INMC, undated.

Aristotle. *The Art of Rhetoric*. London: Penguin Classics, 1991.

Brunsdon, Charlotte. "What Is the 'Television' of Television Studies?". In Horace Newcomb, ed., *Television. The Critical View*, 6th ed., 609–628. New York: Oxford University Press, 2000.

Casey, Bernadette, Neil Casey, Ben Calvert, Liam French and Justin Lewis. *Television Studies. The Key Concepts*. London: Routledge, 2002.

Durkheim, Émile. *Les Formes Élémentaires de la Vie Religieuse*, 5th ed. Paris: PUF, 1968.

Garrett, Almeida. "Ao Conservatório Real, Memória lida em conferência do Conservatório Real de Lisboa em 6 de Maio de 1843." *Obras Completas*, Vol. I. Empresa de História de Portugal da Sociedade Editora, 1904.

Goethe, Johann W. *Fausto*. Lisboa: Relógio d'Água, 1999.

Hartley, John. *Uses of Television*. London: Routledge, 1999.

Liebes, Tamar. "Television's Disaster Marathons: A Danger for Democratic Processes?" In Tamar Liebes and James Curran, eds., *Media, Ritual and Identity*, 71–84. London: Routledge, 1998.

Mauss, Marcel. *Manuel d'Ethnographie*, 2nd ed. Paris: Payot, 2002.

Morin, Edgar. "Une télé-tragédie américaine: l'assassinat du Président Kennedy." *Communications*, no. 3 (1964): 77–81.

Nabokov, Vladimir. "On a Book Entitled *Lolita*". In V. Nabokov, *Lolita*. London: Penguin, 1995.

Nuttall, Anthony David. *Why Does Tragedy Give Pleasure*. Oxford: Clarendon Press, 1996.

Tönnies, Ferdinand. *Comunidad y Asociación*. Barcelona: Edicions 62, 1979.

Torres, Eduardo Cintra. *A Tragedia Televisiva: Um Género Dramático da Informação Audiovisual*. Lisboa: Imprensa de Ciências Sociais, 2006.

Vernant, Jean-Pierre, and Pierre Vidal-Naquet. *Mythe et Tragédie en Grèce Ancienne*, vol. I. Paris: La Découverte, 2001.

Weber, Max. *Economia y Sociedad*. Mexico City: Fondo de Cultura Económica, 1983.

Fernando Ilharco
Screens of Fire: Surviving the End of the World

Growing complexity inevitably leads to catastrophe. Human beings have for long adapted to a chaotic world: language, culture, science and technology are developments appropriately described under this perspective. Either by simplification or description the complexity of a world already there is reduced. The path of information and communication technology, of digital screens, however, might embody a new strategy for surviving: substitution, protection, that is, screening. Human life nowadays happens behind the screens (Introna and Ilharco 2011, 2006; Ilharco 2008).

All philosophies and social theories fundamentally, explicitly or implicitly, deal with the question of death, as referred Schopenhauer; and digital screens also deal with the question of death. TV, computer and mobile phone screens are the focal points of human attention. In the semiotic contemporary culture of abundance, the complexity of the world is back as a message. Ted Turner, the founder of CNN, warned us in the 1980s: "CNN will broadcast live the end of the world." The dominant red color of the CNN logo and indeed of TV images is an attractor that captures attention, an evocation of the epical fires that wrote History. Red has long meant attention, fire, change, accident.

The deeper message of TV, of live TV and the Internet is the final catastrophe that complexity leads to. The digital screen is a showcase of catastrophes. Staring at the images, watching the screens, the viewers are outside the real world, protected by a screen. Living in a screened cultural landscape, watching the catastrophes, viewers are separated, protected, excluded by the screens. The screen is the distinction that draws contemporary times. This side of screens (where we talk and write papers), men experience the feeling of the survivor, living while others die. Immersed in a hyper-real world, a reality made of images, immateriality and change, the screen-watcher is drawn into the final paradoxical show: the end of the world, and surviving it.

1 A Screen of Life

What do screens screen? What does a screen do? Obviously, it screens. Screens what? A screen in screening gathers the attention of the people that surround it. In screening, it acts as the location where what is supposedly relevant will be seen. When we consider the screen, as it appears in its world, it seems to appear as something that calls for or grabs our attention. Without this already calling for our attention, screens would no longer be screens. The description of a screen points to the notions of presenting, making present, gathering attention, suggesting relevance and acting as a medium. These ideas emerge in and through our involvement in particular ongoing activities.

Screens flow along by making evident our involvement in-the-world. They present an already screened world to us and already consistent with our ongoing involvement in that world. As screens, we look at them but also simultaneously, immediately, and more fundamentally, we look through them to encounter our way of being-in-the-world (Heidegger 1962).

Hence, screens are not mirrors in that they do not reflect whatever they face. They are rather surfaces that present what is already relevant within the flow of our purposeful action. However, it must also be noted that in presenting or displaying – in making relevant or evident – other possibilities are simultaneously excluded. This is precisely one of the central common meanings of the word screening today (as in selecting or choosing).

Screening, as inclusion and exclusion, is therefore also a framing process. For these screening surfaces present what is already relevant within the flow of our purposeful action. However, it must also be noted that in presenting the subject dies. This agreement for screening, as in including and excluding, is not about the content of this or that screen but rather an already agreed agreement about a particular way of living or form of life (Wittgenstein 1967). This way of living, that is the implied criteria of agreement, addresses something even more fundamental, namely the realm of truth. Heidegger (1977) noted in his investigation of the Greek concept of truth that the Greek word for truth, *aletheia*, meant the simultaneous revealing and concealing of something.

We might suggest that screens, as focal surfaces that grab and hold our attention, may indeed also appear to us as what is already relevant within the flow of our purposeful action. However, we would again note the death inherent in presentation. As the grounding context of truth, screens condition that which can legitimately be asserted. This represents an important conclusion if we consider the primacy of seeing in the Western way of thinking and living, often expressed through the saying "seeing is believing".

In Heidegger's terminology, we can say that the screen is a kind of Ge-stell, or 'enframing' (Heidegger 1977). The screening of screens is a kind of framework or frame at work, in which the possibilities for truth, our mostly implicitly agreed way of living and doing, are the background that make relevance appear.

2 Screening the Screen

The word screen is both a noun and a verb. Its contemporary plurality of meanings can be collected along three main themes: projecting/showing (TV screen), hiding/protecting (fireplace screen), and testing/selecting (screening the candidates) (OPDT 1997, 681–682).

The origins of the word screen go back to the fourteenth century. According to the Webster Dictionary (1999), the contemporary English word 'screen' evolved from the Middle English word 'screne', from the Middle French 'escren', and from the Middle Dutch 'scherm'. It is a word akin to the Old High German (eighth century) words 'skirm', which meant shield, and 'skrank', which meant a 'barrier' of some kind.

The word 'screen' suggests still another interesting signification, further away from us in history. It is a word "probably akin" (MW 1999) to the Sanskrit (1000 BC) words 'carman', which meant skin, and 'kränti', which signifies 'he injures'. These meanings, possibly, are those from which the Middle Age words evolved. The Sanskrit origins suggest that the notions of protection, shield, barrier, separation, arose as metaphors of the concept of skin, possibly human or animal skin.

This etymological analysis indicates that the word 'screen' moved from the Sanskrit meaning of skin and injury, through protecting, sheltering and covering, to the modern day projecting, showing, revealing, as well as electing, detecting and testing. Now, we may ask the following: is there any central intent, distinction or feature common to all these specific meanings of the word screen? We believe the answer is yes; after all, that is the same word. To defend such an assertion, we take up a different but related route: that of the analysis of sound, a practice known as sound symbolism or phonosemantics (e.g., Jakobson and Waugh 2002; Magnus 1999).

The word 'screen' is pronounced 'skri:n': very close in its sound to the word 'scream', pronounced 'skri:m'. It is just a final sound that distinguishes both words. The core sound of both words is the same – skri. Do both point to something beyond themselves? Does that common sound have a meaning of its own?

The correspondence between sounds and meanings remains an enigma to a great extent. One should remember that in 1866 the Linguistic Society of Paris banned discussions on the origins of language. That prohibition remained influential across much of the western world until the second half of the twentieth century (Stam 1976); and, indeed, in a way, still in effect. Thus, the relationship between meanings and sounds has for a long time proven an issue that approaches the contours of a taboo. Yet, one would say in a sensible way that the issue is pertinent because there are simply too many clues, coincidences and indications for the question to be meaningless. Nevertheless, researchers in the field have thus far been incapable of progressing in a recognized way in this manifestly difficult area.

However, this state of affairs does not mean that this kind of sound analysis is senseless. Quite the contrary, in some cases "speakers feel (that certain forms in language) do have a close relationship to objects or states in the outside world [means that] individual sounds are thought to reflect, or symbolise, properties of the world, and thus to 'have meaning' " (Crystal 1987, 174).

When we look carefully at what those words mean in this kind of sound analysis in different languages, we can discern some interesting insights. In other languages, such as French (*cri*), Czech (*vykrik*), Danish (*skrig*), German (*schrei*) or Italian – *grido* or *urlo*, which is close to the Portuguese *urro*, meaning roar –, and others would be of use as well. The Portuguese word for screen is '*écrã*' (pronounced '*Ekrã*') and for scream it is '*grito*' (pronounced '*gri'tu*'). Quite different words at first glance. However, these two Portuguese words, as is the case for the two English words referred to above, have a common core sound: 'kr' (voiceless plosive) and 'gr' (voiced plosive) are the same sound but for a very minor variation. Thus, a question one needs to consider is whether there is a certain onomatopoeia in that sound, in both words, which evokes some particular meaning. Our suggestion, strictly an exploratory one, is that the sound gr/kr has an onomatopoeian meaning that refers to the roaring of animals; this deep reference, in its turn, refers to a deeper meaning of attention.

By now, taking into account the analysis so far performed on 'screen', and the common meaning of 'scream', the answer appears intuitively: 'to call for attention'. 'To call for attention' seems to be the meaning attached to the sound 'kr/kri' for both words in both languages. The animal clue in the scream, in the Italian *urlo* is indeed intriguing; what does gr mean? What gr, as a scream, perhaps grrrrrrr means? Perhaps a roaring. Roaring has long since been a call for attention for humans. As McLuhan (1994) refers, long before the alphabet, that introduced the primacy of vision, "hearing was believing". Furthermore, it is quite elucidative that for many years one of the major Hollywood movies corporations, MGM, opened its movie sessions with the roaring of a lion ... attention, the movie, action on the screen, is about to begin.

A further scrutiny of the sounds in question strengthens this path. The Portuguese word 'écrã', a quite recent word coming from the French language, is close to the Middle French word 'escren', referred to above. The corresponding actual French word is 'écran'. Is it the case that this initial 'e' – sound "e" –, which the word screen does not have, points to a specific, old, primordial meaning?

On first consideration, the meaning of the 'e' may seem a worthless question as it is a widely recognized principle of phonetics "that individual sounds do not have meaning: it does not make sense to ask what 'p' or 'a' [or 'e'] mean" (Crystal 1987, 174).

The 'e' we are addressing is the sound ''e'. Thus, the appropriate question is: "what does this ''e' sound mean?" The answer is an intriguing one. The letter 'e', that represents the sound in question, is widely used in Portuguese as a prefix. Yet, as we further inquire into the sound ''e', we note there is indeed a Portuguese word that only has that sound: the word "eh!" (DLP 1989, 580). This word is a grammatical interjection, a "designate of surprise, admiration, and calling" (580; our translation). Of course, we are referring to words on the one hand and sounds on the other hand. However, the widely accepted position on the arbitrariness of sounds is more dogma than evidence. Because language, being what it is, obviously gets shared, its path most possibly was – as is – one of conventionalization. Hundreds of thousands or millions of years have gone by and it would be sensible to accept the loss of clues on the dawn of language, humanity, civilization. In fact, it is not an argument of ours that sounds do have relationships to meaning; in dense, complex, intricate ways of course; perhaps difficult or even impossible to distinguish. There is a whole tradition on this approach. From Plato's *Cratylus* (Plato 2008) to Merritt Ruhlen (1997), Corballis (2003) and along with other contemporary works, many authors do acknowledge intriguing connections between sound and meaning. Thus, recalling the necessary falsifiability of scientific positions, we think it appropriate to leave a door open on some of the enigmatic clues we point out in this paper.

The Portuguese word 'eh!' commonly serves in situations where someone wants to call the attention of someone else. Let us consider an example and the respective English translation: "Eh! Anda cá" means in English "Hey! Come here." We evidently note that in both languages the first word 'eh'/'hey' is pronounced, that is, is destined, to call the attention of the one we want to "come here". The English word 'hey', the most common translation of the Portuguese word 'eh' (MID 1958, 502), means exactly "calling attention" (OPDT 1997, 349).

If this is so, one may ask what about the word 'screen', it has no ''e' sound? Maybe the word screen never had an initial ''e' because the English language, as it evolved from the Middle Ages to the present day, seems to have encountered other solutions to emphasize the meanings at stake: the sounds 'scr' and 'i' instead of the sound 'é' as in Portuguese, French and other languages.

Thus, we would conclude that from the Middle Ages to the present day the evolution of the Middle French word 'escren' and that of the Middle English word 'screne' seem to have followed the same path. Escren became écran, screne became screen, that is, in both evolutions, each according to its own context, the words moved towards stressing the 'call for attention'. Screen is attention. It essentially is attention.

3 A Screen against the End

In a screened world, in an attention society, in a cultural space where everything is calling our attention, that is, where attention is the scarcest resource, catastrophes are the most promising spectacle, the call for attention – red, flames, fire.

In hyper-reality, in the third order of simulacra (Baudrillard 2004), fires on television are just information. We question: what does the fire on the screens inform us about?

Fire is both the symbol that always has been and a medium of a screened hyper-reality, contextualized by abundance, the world of mobile phones, MTV, the Internet and advertising. Fires on television screens are both cause and consequence of the hyper-real regime in which we live, as well as a paradoxical symbolic exchange that re-feeds this same hyper-real framework.

Calling for attention, stressing red, fires on the screen fuels the drama. The drama on the screens is essentially the drama of the announcement of the end. All the rest, everything on the surface, are just details, some kind of a communicational varnish, a skin that covers reality – a screen over the real, a screen against the world, against the end.

It is suspected from the beginning of the century – what is all this, airplanes, embassies, trains, cars, flags, people on fire every day in our home? "The end will be broadcast live", Ted Turner told us 30 years ago, when he founded CNN. It is no coincidence that the key color of CNN is red. All that a TV channel broadcasts nowadays is a second choice, a minor variation, compared to the first and most desirable issue of all: the last show, the end of times.

The 'end', promised by Ted Turner, is a word on the cover of dozens of bestsellers; from humanity, mortality and work, to the state, science and many other subjects. The end was early on in our epoch staged on the beaches of Vietnam, rearticulating history under the sound of Wagner, The Doors and the fire of napalm. If indeed America lost that war – Vietnam is today an emerging market economy – the United States won a global industry, the movies, the screens, the hyper-real. More, they led reality as screen. As Baudrillard said: "we enter our life as we walk into a screen."

In hyper-reality, the signifier is always a sign. Reality is semiotics. And nature is a strange fiction. For us, the hostages of the screen (Baudrillard 2002), slaves of advertising, television and the Internet, prisoners of the third order of simulacra, the final myth that establishes a sense of substance is found in the thousands of cars torched in Paris in November 2005, in the bombs on trains in the centre of Madrid, in the embassies on fire in the cartoon war, in the giant summer fires in Portuguese forests, in the fires on the television screens. This myth is the Armageddon, the ultimate and monstrous battle between good and evil, surrounded by the fires of hell. "Witnesses of universal apocalyptic events", in the words of Habermas (2003, 101–102) referring to 9/11, "we are assailed by biblical images as we watch television again and again, something a masochistic attitude, showing the images of the towers of Manhattan to crumble. The very language of retaliation has an aura of the Old Testament."

The power of fire is the power of history. The power of fire is the power of real time, creation, the future, transformation. The control of fire by man ultimately is an extension of the creative act, the generation of whatever arises – the most powerful media in the history of mankind, as McLuhan referred to it. Fire is a strange attractor, drawing, manipulating, attracting media attention because, burning, destroying, obliterating, murdering, exterminating, and overturning fire brings fear, anxiety and terror to anyone but the media, whom it brings audiences.

Being in itself change, fire is used as seduction by the media. The message of fire on television screens is the end of times, here and now. Fire absorbs everything as if some black hole. On the screen, we watch it.

Fire is the power of change, it is change, right in front of us, recalling Heraclitus (c. 540 BC – c. 480 BC). The final change, the deeper meaning of fire, is the essence of television, of the media. For a long time, the media, because of its immense power of destruction, are identified with the forces of evil. The British magazine *The Economist* considered the Internet as the most powerful weapon of new terrorism ... and that is so; so much so that we would find that the most effective strategy to end global terrorism would be simply to close the Internet, and CNN ...

Perhaps the ancient belief that fire is a messenger between the worlds of the living and the dead is accurate. After all, it is no coincidence that weapons are precisely firearms.

Fire is the color of change. It is the color of blood. Revolutions are made with firearms. Red is for stop. Red is for fear and for protection. It is the highest alert level color of the alert scale anti-terrorist of the U.S. Homeland Security Department, indicating the near certainty of a terrorist attack and its high gravity. "Red means run son, numbers add up to nothing", sings Neil Young. Cause and consequence, red is the color of fire. Again, it is no coincidence that the colors of CNN are the colors of fire: red, orange, black, gray and their many variations. In general, in the Western world red means the most danger. Emergency exits are red. The traffic lights are to stop. In a race car, the red flag means that all cars must stop immediately. In the global game of football, the red card sees the expulsion of the player. In exact sciences, the 'red line' means the maximum that can be developed in certain operations.

This symbolism goes back to the mythology of imperial Rome when red was associated with the god of war, Mars; Mars today, the red planet. In the science fiction television series *Star Trek*, the security professionals, who regularly suffer casualties in outside missions, wear red uniforms. In the film and television industries, the redshirt means a character-type whose main feature is to die violently soon after being introduced into the story.

The colors of fire are the colors of catastrophe, the colors of news because all are changed. With flames on TV screens, it is as the fire of Heraclitus were back for revenge, because we have taken the side of Parmenides, the repetition, the cause and effect assumption upon which science is based.

News is what changes the course of the day, people's lives and times and catastrophe is the biggest news of all. Thus, red, the color of fire, blood, life, strength, power and attention, is also the color of news. The contemporary media world is now a supercool environment, exerting upon us a tremendous fascination and seduction, which is not easy to resist. It is precisely this supercool seduction, advertising, MTV, millions of dollars, rock music, mobile phones, cars, the fashion of clothing, of instant millionaires, that at any moment can turn into fire – in the Manhattan fire towers, in the heart of Paris, in the luxury resorts of Bali, or hidden away, deep in the tunnels of the London Underground.

4 The Final Screen

All these meanings, literally and metaphorically, are relevant to trying to better understand how to describe fire and red on screens – growing in the streets of Europe and America, as the show central to television; the riots of November 2005 in Paris, in the ongoing massive summer forest fires in Portugal over the last decade, in Madrid in the train bombs of 11 March, in the embassies in the cartoon war, and in the most strange of all events of our era, the blasting of two U.S. planes into the twin towers of Manhattan. The fire provides access to the screen.

The background narrative that is part of the apocalyptic destruction of the hurricane Katrina, global terrorism, the war in Iraq, the explosion of the space shuttle, bird flu, all are global crises of control systems, the emergence of an empire, not of Washington but of accidents. It is the final catastrophe that is the essence of global television. In an instant, a city was destroyed. Not just war on the screens, but in the promised land, in the heart of the new Rome. In a second, suddenly, a massive crash took the world's attention. In our era, where "television is the museum of accidents" (Virilio 1994), the successes and failures, the difference that counts, reaches us by surprise. This pattern of immediacy is related to the speed of electronic technologies as well as to the way nature builds tensions and suddenly releases them, purging impurities and excess.

On the one hand, it is easy to conclude that 11 September, Iraq, Katrina, the London bombings, 11 March in Madrid, the Portuguese summer fires etc. weaken the systems of power, of control. On the other hand, from a systemic point of view, taking long-term survival as the ultimate criterion, any failure as long as not being fatal, is an enhancement, an additional motivation – "what does not kill me, makes me stronger", wrote Nietzsche. But, as Joker, the bad guy of Batman the movie tells us: "What does not kill me, makes me stranger ..." Yet, obviously, what kills, really kills. What destroys, destroys. Or as Baudrillard writes (2005, 193), "too much is too much."

Nonetheless, as Baudrillard (2002) advised us, all catastrophes protect us from something worse. He refers to neurosis, which protects us from the most complete madness, or AIDS, which protects us from the most complete promiscuity.

What, then, can catastrophes on the screens protect us from? As neurosis or AIDS protect us against the excesses they prevent, the final catastrophe in the media, the promise of Ted Turner, protects us from a major catastrophe. What may be even a greater catastrophe than the screened final show? Surely, material and symbolical, it would be the final catastrophe in the real world. The end of the world on the television, making us play the role of the survivor, is a final warning. It is a screen, a shield, a protection against the end of the world, the Armageddon, not on screens but in the real world.

Remembering Marcel Proust's final twist in *À la Recherche du Temps Perdu*, after re-living, capturing all his life, Marcel begins the writing of the novel we are then reading. Just as Proust, doing justice to Ted Turner's prophecy, the end would come on television. As with Saddam Hussein, Milošević, Bin Laden or Gadaffi, the screen watcher will see the final shadow looming over him as on television, indeed on television. Deeply involved, the end would be all around, he will be in the centre, watched by all the screens of the world, and then it will be the end.

References

Baudrillard, Jean. *Screened Out*. London and New York: Verso, 2002 (2000).

Baudrillard, Jean. *Simulacra and Simulation*. Chicago: The University of Chicago Press, 2004 (1981).

Baudrillard, Jean. *The Intelligence of Evil or The Lucidity Pact*. Oxford and New York: Berg, 2005.

Corballis, Michael C. *From Hand to Mouth: The Origins of Language*. Princeton: Princeton University Press, 2003.

Crystal, David. *The Cambridge Encyclopedia of Language*. Cambridge, New York, Melbourne and Sydney: Cambridge University Press, 1987.

DLP. *Dicionário da Língua Portuguesa*. Porto: Porto Editora, 1989.

Habermas, Jürgen. *The Future of Human Nature*. London: Polity, 2003.

Heidegger, Martin. *Being and Time*. Trans. John Macquarrie, and Edward Robinson. Oxford and Cambridge, MA: Blackwell, 1962.

Heidegger, Martin. *The Question Concerning Technology and Other Essays*. New York: Harper Torchbooks, 1977.

Ilharco, Fernando M. "A Catarse do Fogo: A Simbologia do Fogo nos Ecrãs da Televisão". *Comunicação & Cultura* no. 5 (2008): 139–153.

Introna, Lucas D., and Fernando M. Ilharco. "On the Meaning of Screens: Towards a Phenomenological Account of *Screenness*." Human Studies 29, no. 1 (2006): 57–67.

Introna, Lucas D., and Fernando M. Ilharco. "Phenomenology, Srens, and 'Screenness': Returning to the World Itself." In *The Oxford Handbook of Management Information Systems. Critical Perspectives and New Directions,* ed. Robert D. Galliers and Wendy L. Currie, 253–292. Oxford: Oxford University Press, 2011.

Jakobson, Roman, and Linda R. Waugh. *The Sound Shape of Language*. 3rd edition. Berlin: De Gruyter Mouton, 2002 (1979).

Magnus, Margaret. *Gods of the Word: Archetypes in the Consonants*. Philadelphia: Thomas Jefferson University Press, 1999.

McLuhan, Marshall. *Understanding Media*. Cambridge, MA: MIT Press, 1994 (1964).

MID. *Michaelis Illustrated Dictionary*. Vol. I. English-Portuguese. São Paulo: Ediçøes Melhoramentos, 1958.

OPDT. *Oxford Paperback Dictionary & Thesaurus*. Ed. Julia Elliot. Oxford and New York: Oxford University Press, 1997.

Plato. *Cratylus*. London: Arc Manor, 2008.

Ruhlen, Merritt. *On the Origin of Languages*. Stanford: Stanford University Press, 1997.

Stam, James H. *Inquiries into the Origins of Language*. New York: Harper and Row, 1976.

Virilio, Paul. *The Vision Machine*. Bloomington and London: Indiana University Press and British Film Institute, 1994 (1988).

Wittgenstein, Ludwig. *Philosophical Investigations*. Oxford: Blackwell, 1967.

Internet:

MW (1999) *Merriam-Webster Dictionary* (2010). http://www.m-w.com.

Isak Winkel Holm
Zombies and Citizens

The Ontopolitics of Disaster in Francis Lawrence's *I Am Legend*

"In this case the measles, um, virus which has been engineered at a genetic level to be helpful rather than harmful. Um, I find the best way to describe it is if you can … if you can imagine your body as a highway, and you picture the virus as a very fast car, um, being driven by a very bad man. Imagine the damage that car can cause. Then if you replace that man with a cop … the picture changes. And that's essentially what we've done." In the first scene of Francis Lawrence's movie *I Am Legend* from 2007, Dr Alice Krippin, played by Emma Thompson, proudly explains her cancer research to the TV viewers. "So you have actually cured cancer?" the journalist asks; "Yes, yes … yes, we have." The next scene shows a deserted New York City three years later where grass has begun to sprout in the streets. What the research team has done, essentially, is creating a plague that has wiped out 90% of the human race and made the rest mutate into so-called Darkseekers, speechless monsters allergic to daylight and filled with cannibalistic rage.

Suddenly a very fast sports car comes racing down an empty avenue. Dr Krippin's figure of speech has turned into an objective fact, and this poses a question about the driver, the immune Robert Neville, played by Will Smith: is he a cop or a very bad man? Prior to the virus apocalypse, Dr Neville seems to have had some kind of medical job in the army, and in the post-apocalyptic New York, he uses his mixture of medical and military skills in his efforts to save the human race: he catches the Darkseekers, uses them as test animals in his basement laboratory, and kills them afterwards. The question is whether this biopolitical violence is helpful or harmful.

The Darkseekers are mutants, not remnants from the dead, and they move around like top athletes, not in the famous slow-motion stagger. But even if the monsters cannot be categorized as regular zombies, the movie unfolds in what I suggest calling a zombified space: a social space divided into two kinds of human beings with unequal ontological and legal status. It is the zombification of the post-plague New York City that turns Dr Krippin's slightly infantile distinction between the cop and the "very bad man" into the fundamental question of the movie. If the chasm cutting through social space is so deep that the Darkseekers do not have any human features at all, it could be claimed that the violence Neville uses is justified. Nevertheless, if the zombification is not complete and they still qualify as some kind of fellow human beings with a claim to justice, his violence is criminal.

The modern zombie movie emerged with George A. Romero's *Night of the Living Dead* from 1968. Novel and movie zombies also featured before 1968 but Romero transplanted the zombie motif from the gothic genre to the genre of the disaster movie. The modern zombie movie is, basically, a sub-genre of the disaster movie. According to a survey, zombie apocalypses happen to be the most frequent kind of disaster in the post-apocalyptic movies of the last decade (Drezner 2011, 2). Recent scholars conceptualize the intimate relation between zombies and disasters by pointing out the state of emergency as a central feature of modern zombie movies (Bishop 2010, 11; Canavan 2011, 175; Lillemose and Meyhoff 2011; Brooks 2003, 155). In *Night of the Living Dead*, for instance, before the current is cut, a TV speaker describes the situation as an "emergency". The centrality of the state of emergency in the zombie movie is evident and serves to explain the zombie movie's remarkable rise from the dead in the decade of the so-called war against terror following 9/11.

The problem, however, stems from how most scholars apply the concept of the state of emergency (or its synonym, the state of exception) as a catchphrase for the total breakdown of social order. In the first pages of *Political Theology*, Carl Schmitt underlines that "exception is different from anarchy and chaos" (Schmitt 1985, 12). As Schmitt writes, the state of emergency is not a collapse of all societal infrastructures but a conflict between the law and the medium of law. Medium, in this case, is not the language in which law is articulated but the social reality to which law is applied: "Every general norm demands a normal, everyday frame of life to which it can be factually applied and which is subjected to its regulations. The norm requires a homogeneous medium" (13). Zombified space is a heterogeneous medium in which the ontopolitical incongruity between citizens and zombies makes it impossible to apply the general norms of law. When zombies flood the streets, the "normal, everyday frame of life" breaks down (in German: the normal *Gestaltung der Lebensverhältnisse*, that is, the normal shape or formation of living conditions; Schmitt 1934, 19).

As Colson Whitehead writes in his remarkable zombie novel from 2011, the zombies, despite their wounds and panoply of leaking orifices, still proudly indicate the social tribes to which they had belonged thanks to their "gray pinstriped suits, classic rock T-shirts, cowboy boots, dashikis, striped cashmere cardigans, fringed suede vests, plush jogging suits" (Whitehead 2012, 243). Even though the zombies still display their former life as consumers, their lack of volitional and rational faculties mean that they have lost their life as legal and political persons. Thus, the question is whether the heterogeneity of social space is so grave that one has, at least for a while, to suspend law in order to fix its medium – by putting down the ghouls with the classical headshot. According to Schmitt, this is the kind of question to which the sovereign supplies an answer: "Sovereign is he who decides on the exception" (Schmitt 1985, 5). In Schmittian terms, then, the zombification of social space hinges on a sovereign decision that draws a line between those who count as legal persons and those who count as legally and politically dead.

In the following pages, I want to explore how the sovereign decision distinguishes between those inside and those outside its jurisdiction. Carl Schmitt and Giorgio Agamben cannot help me in this exploration because of their "decisionist" and "exceptionalist" version of the state of emergency. To them, the sovereign decision is the political equivalent of the miracle: a single, inexplicable moment that takes place in a normless void outside the legal order. Fortunately, recent critics of Schmitt, and, consequently, of Agamben, have argued that the sovereign decision is much more "littered, layered, and complex than that."[1] The social order does not have a single main switch that the sovereign can push (This is, by the way, why "state of emergency" is a more appropriate term than "state of exception.").

Inspired by the criticism of decisionism and exceptionalism, I want to convey how the sovereign decision in zombie apocalypses does not come out of the blue. On the contrary, the decision is supported on a preceding and pre-verbal interpretation process that determines whether a certain kind of human being fits into the normal *Gestaltung* of the living conditions or whether it poses a threat to the social order. Like any other interpretation process, this ontopolitical interpretation is guided by pre-understandings, in this case a repertoire of concepts, images and stories about social life. With Charles Taylor, I suggest calling these socially shared ways in which social spaces are imagined a *social imaginary* (Taylor 2004, 23). Lawrence's *I Am Legend* is remarkably explicit in the way it exposes the social imaginaries that underlie the splitting of the social space into two opposing halves.

My argument proceeds in five steps. First, I will outline a theoretical framework that I hope proves helpful (not harmful) in conceptualizing the role played by social imaginaries in the zombification of social space. Second, I will sketch out the historical background for the zombified space in *I Am Legend*, exploring a similar chasm between zombies and citizens in earlier American disaster fiction, even if there are no actual zombies on the cast list. Third, I will present an analysis of *I Am Legend* focusing on the sovereign decision that draws a line

---

1 I am quoting from Connolly 2005. See also the discussions of the "exceptionalist" version of the state of emergency in Honig 2009, Feldman 2010 and Lazar 2009.

between humans and Darkseekers. Fourth, I will contrast my methodological approach to the zombie movie with other approaches. Fifth, and finally, I will ask whether Lawrence's movie challenges or confirms the zombification of social space.

1 Expository Devices

In *A Theory of Justice*, John Rawls suggests a thought experiment in the classical genre of the state of nature and the social contract. In the "original position", future citizens meet outside society to develop the principles of justice. Since the contingencies making citizens socially advantaged or disadvantaged still remain hidden behind a "veil of ignorance", the parties meeting in this original position are able to enter into an unbiased and rational discussion of the basic structure of society. They are unable to feather their own nest as they do yet not know which nest is theirs. According to Rawls, this theoretical fiction functions like an "expository device" that helps us in analyzing and discussing the basic structure of society (Rawls 1999, 19).

Since the 1755 Lisbon earthquake, fiction writers have adopted disasters as a similar thought experiment. "This is Ground Zero. This is my site. I can fix. I can fix this," Neville claims when the virus disaster unfurls. In the terms of political theory, the Ground Zero of modern disaster fiction represents a close relative of the original position of modern philosophical prose. State of emergency and state of nature are two analogous versions of a generic space outside society where it becomes evident what you have to do to fix society from scratch. In other words, I am suggesting a functional similarity between catastrophism and contractualism. However, there is one crucial difference between using the state of nature and the state of emergency as an expository device. In Rawls' original position, the parties assemble to discuss *how* to allocate social goods between them, but they do not discuss *who* counts as a relevant addressee for the distribution of goods. Rawls conceives society "for the time being as a closed system isolated from other societies" (Rawls 1999, 7, 229). Nobody is moving in or out of this society and nobody is caught in a liminal state between inside and outside. To keep matters brief, Rawls' version of the expository device fails to expose the question of who is present in the original position.

By contrast, the Ground Zero of disaster fiction questions neither the substance nor the procedure of justice but, first of all, the scope of justice. In Hannah Arendt's famous sentence, the problem is not the specific rights of citizens but the basic right to have rights (Arendt 1979, 296–297).

In *Spheres of Justice* (1983), Michael Walzer formulated an important critique of distributive justice that has served as a matrix for philosophical discussions of justice from Aristotle to Rawls. According to Walzer, there is a kind of justice which is not about the distribution of social goods among the members of a society but rather about the distribution of membership itself: "When we think about distributive justice [...] we assume an established group and a fixed population, and so we miss the first and most important question: who constitutes that group?" (Walzer 1983, 31). Since Walzer's book, political philosophers have begun to pay attention to the first and most important question about the scope of justice. The task has been to figure out what justice means if "not just the 'what' but also the 'who' is up for grabs," as Nancy Fraser writes in a recent book (Fraser 2008, 15). If you want the concept of justice to account also for the constitution of the group, it swells to a much broader concept that not only covers the legal rules and rational principles governing the distribution of social goods. Formulated with Carl Schmitt – who, until recently, has played a modest role in this tradition – the question of the scope of justice has shifted the theoretical focus from the norm of law to the medium of law: it is not *Gesetz* but *Gestaltung*, not the legal norms but the basic formation of the living conditions that determine who counts as a subject of justice.[2]

Judith Butler discusses a similar issue in her latest book: "Part of the very problem of contemporary political life is that not everyone counts as a subject," she asserts in *Frames of War* (Butler 2009, 31). Butler's account is useful for a conceptualization of the zombified space of modern disaster fiction for two reasons.

First, Butler's analysis of contemporary political landscape is guided by the concept of the state of emergency. Her examples are the detainees at Guantanamo Bay and other illegal combatants in the global war on terror who, in the West, are not perceived as worthy of protection or even as worthy of grief. Like Walzer, Butler asks how the group of right-bearing subjects is constituted, but in the global perspective of the war on terror, the question of the scope of justice is not simply a matter of membership or non-membership of a nation state. On the other hand, she does not, as Agamben does, make a sweeping claim about the state of exception as the hidden matrix and *nomos* of modern societies

---

2 In recent theory of justice, you can find different approaches to conceptualizing what Schmitt hints at with his idea of a *Gestaltung* of living conditions. Charles Taylor refers to the medium of law as a "framework", a "background against which the principles of distributive justice must operate". According to Taylor, this framework, which he later calls a social imaginary, determines who counts as a "subject to whom distributive justice is due" (Taylor 1985, 295).

(Agamben 1998, 167). To Butler, local states of emergency create a perforated and heterogeneous normative order in which "certain lives are regarded worthy of protection while others are not, precisely because they are not quite 'lives' " (Butler 2009, 50).

Second, Butler focuses on the processes that draw the line between those inside and those outside the political fabric of life. Unconcerned with the legal procedures of citizenship, Butler explores the cultural processes taking place in the social imaginary: "How do the norms that govern which lives will be regarded as human enter into the frames through which discourse and visual representation proceed, and how do these in turn delimit or orchestrate our ethical responsiveness to suffering?" (77). In Butler's vocabulary, a frame is a mode of public seeing (Butler 2004, 147) or a mode of intelligibility (Butler 2009, 149) that determines how different kinds of human beings enter into "the realm of appearance" (140).

According to Butler, the politics of rational rules and principles is based on a politics at an "ontological level" (138). In this context, "ontology" is not a question of what human beings actually *are* but of how they *appear to be* to the public mind. In William Connolly's terms, Butler's central argument is that our ethical responsiveness to other human beings is orchestrated on an ontopolitical level, not on the level of formulated norms.[3]

With Butler's own example, the pictures the US Department of Defense published from the Detention Camp at Guantanamo Bay framed the detainees as non-human creatures. In this case, the framing was a dense network of metaphorical *as if*'s transforming the detainees into faceless and abject non-humans: in their chicken cages, with manacled hands and goggles, they were represented as if they were caged animals, as if they were mentally ill, or even as if they were dead (Butler 2004, 72).

One could summarize Butler's theory of the dehumanizing frames in Carl Schmitt's terms. According to Schmitt, the state of emergency is a situation in which the heterogeneous medium of the law blocks the application of legal norms. In Butler's argument, the Schmittian distinction between the law and the medium of the law is mirrored in a distinction between the normative and the descriptive: on the one hand, the compendium of moral and legal norms, on the other hand, the definition of specific phenomena. Normally, we understand

---

3 "*Onto*, because every political interpretation invokes a set of fundaments about necessities and possibilities of human being, about, for instance, the forms into which humans may be composed and the possible relations humans can establish with nature. An ontopolitical stance, for instance, might strive to articulate a law or design set into the very order of things" (Connolly 1995, 1).

a judgment as an application of a moral or legal norm to a neutral phenomenon. The point that Butler stresses, however, is that the distinction between the descriptive and the normative is confused since the norms – understood as the frames governing the linguistic and visual representation of the phenomenon – are already at work at the descriptive level formatting the ontological status of the phenomenon: "the judgment is built into the definition (we are, in fact, judging before knowing)" (Butler 2009, 155). Even before we start thinking about the rights of the detainees at Guantanamo Bay, we perceive them as less than human. Butler teaches us that the zombification of social space is not just due to the objective differences between the creatures but, first of all, by the "differential operations" of a social imaginary (77).

2 Voight-Kampff Devices

"My name is Robert Neville. I am a survivor living in New York City. I am broadcasting on all AM frequencies. I will be at the South Street Seaport everyday at mid-day, when the sun is highest in the sky. If you are out there … if anyone is out there … I can provide food, I can provide shelter, I can provide security. If there's anybody out there … anybody … please. You are not alone." In his daily radio broadcast, Neville implicitly draws a line between those he will provide protection for and those he will provide protection against. He is willing to help his fellow human beings to whom he refers to as "anyone" and "anybody", but he is unwilling to protect the Darkseekers since they are not included in the pronouns "anyone" and "anybody".

Had the question been how Neville and the Darkseekers should distribute food, shelter and protection among themselves, it would have been a classical problem of distributive justice. Instead, Neville's radio message poses a question of the scope of justice: do the Darkseekers count as members of the human community? Do they belong to the circle of those entitled to a share of the social goods? The negative answer to this question is not given by Neville's sovereign decision but, rather, by his preconscious interpretation of the Darkseekers as humanoid beings that cannot be included in the pronouns "anyone" and "anybody". In zombified space, the "veil of ignorance" falls so deep that it hides not only the Darkseekers' social status in a future society but also their ontological status as human beings.

The question of the scope of justice is also present in the works of disaster fiction that inspired *I Am Legend*. The most obvious precursor of the movie is, of course, Richard Matheson's novel *I am Legend* (1954), of which Lawrence's movie is the third adaption so far. In Matheson's original version of the story, a nuclear war has turned the human race into vampires. Robert Neville, who happens to be immune, spends his lonely days in Los Angeles by driving spikes through the bodies of sleeping vampires. More than once, however, he feels an urge to justify his acts of violence: " 'You can't abide by Robert's Rules of Order in the jungle,' he said. 'Believe me, it's the only thing I can do. Is it better to let them die of the disease and return – in a far more terrible way?' " (Matheson 1999, 136). In this case Robert's Rules of Order, a codification of American parliamentary law, function as a synecdoche for justice as such. It should be noted that Neville, unconcerned with the substance of these rules of order, poses the question of their scope. The monsters simply do not show up on his radar of justice. In the vocabulary of Carl Schmitt, the post-apocalyptic "jungle" is not the kind of homogeneous medium required by the norm of law. By the way, George A. Romero has admitted that *Night of the Living Dead* basically was a rip-off of Matheson's novel (Kay 2008, V); the structural similarity between Romero's zombie apocalypse and Lawrence's virus disaster is, to some degree, a matter of common sources.

Another important precursor is Ridley Scott's *Blade Runner* (1982), based on Philip K. Dick's novel *Do Androids Dream of Electric Sheep* (1968). In fact, Scott worked as co-writer on the early drafts of the script of *I Am Legend* before Warner Bros. dismissed him and hired Lawrence. In the bleak post-apocalyptic world of *Blade Runner*, after a nuclear war has wiped out most animals of the planet, the pressing problem seems to be drawing the line between humans and genetically engineered organic robots. For that purpose, the protagonist Rick Deckard repeatedly runs the so-called Voight-Kampff device to make sure that the test person is really a human being and not a replicant (or an android, as the fake humans are called in the novel). The interpretive tool measures respiration, "blush response", heart rate, and eye movement in response to emotionally provocative questions. If no human empathy is measured, the test person is shot dead on the spot. The Voight-Kampff device is a modern version of the emblematic balance of justice – with the important difference that it measures not the substance but the scope of justice.

Apart from great images of deserted cities and crumbling infrastructure, the most interesting thing about modern disaster fiction is that it exposes the process of drawing a line between those inside and those outside the jurisdiction. Fictional disasters have a tendency to create what I call a zombified space: they do not just decimate the human race, they also divide it into humans and non-humans. Already in Heinrich von Kleist's short story "The Earthquake in Chile" from 1806, one of the founding texts of modern disaster fiction to emerge out of the rubble of the Lisbon earthquake in 1755, the disaster and the ensuing state of emergency split the survivors into members and non-members of Chilean society (Holm 2012). Even if there are no zombies on the role list, the disaster creates a "zombified" space. By exposing the zombification of social space, modern disaster fiction demonstrates that sovereign decisions do not hover in a vacuous and miraculous void. Instead, sovereign decisions are based on sovereign interpretations, hermeneutic practices that take place in a dense cultural space. A sovereign will always be in need of some kind of Voight-Kampff device.

3 Framing the Darkseekers

"Behavioural note – an infected male exposed himself to sunlight today. Now it's possible decreased function or growing scarcity of food is causing them to … ignore their basic survival instincts. Social de-evolution appears complete. Typical human behaviour is now entirely absent." This note, dictated by Neville to his computer after having observed a Darkseeker behaving strangely earlier the same day, is an interpretation. He starts out from a specific detail – the Darkseeker's exposing itself to the sunlight to which it is allergic – and moves from there to a general interpretation of all Darkseekers as completely dehumanized. Since Neville is sitting in his basement laboratory dressed in a white coat, it seems as if he bases his interpretation of the Darkseekers on their objective behavioral features. However, his judgment is already built into his linguistic description of the Darkseeker as "an infected male" (as opposed to "an infected man"). Later on in the movie, he refers to a female Darkseeker as "it" (as opposed to "she"). Even before the scientific interpretation, Neville's use of language has excluded the Darkseeker from the human pronouns "anyone" and "anybody". Hence, the zombification is not just a physiological but also a cultural process. Neville's sovereign decision about the Darkseekers' non-human status is based on a preceding and pre-conscious interpretation in which his social imaginary frames the ontopolitical status of infected. They are monsters because of a rhetorical mutation and not just because of a genetic mutation.

My first example of this rhetorical mutation is a metaphor. Before Neville dictates his behavioral note to his computer, in a flashback at the very beginning of the movie, he is driving his car, trying to get his wife and daughter out of New York City before the state applies its biopolitical power and quarantines the city. Suddenly, the first Darkseeker of the movie hits the windshield in a short, nearly subliminal image. (The scene is, by the way, a quote of the first zombie attacking the car of the female protagonist at the beginning of Romero's *Night of the Living Dead*). In the next scene, Neville wakes up alone in his fortified home at Washington Square Park and goes down to his basement laboratory to check the infected rats on which he has tried different samples of cure for the virus. The rats are kept in small cages and when Neville draws aside a curtain, a rage-infected rat attacks and slams against a glass screen.

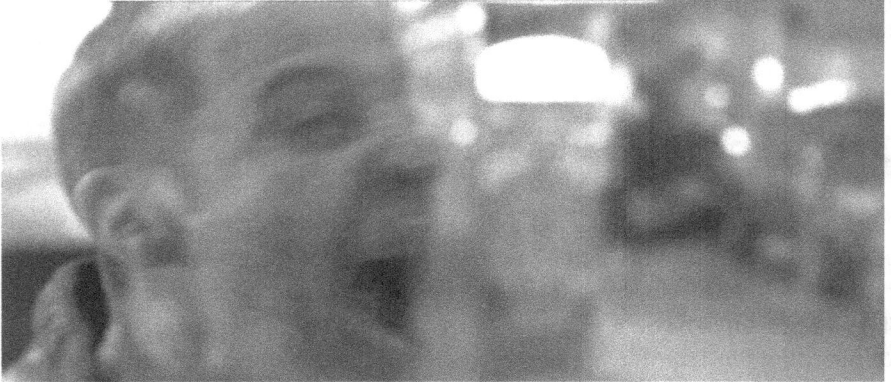

Fig. 1: Darkseeker smashing into windshield

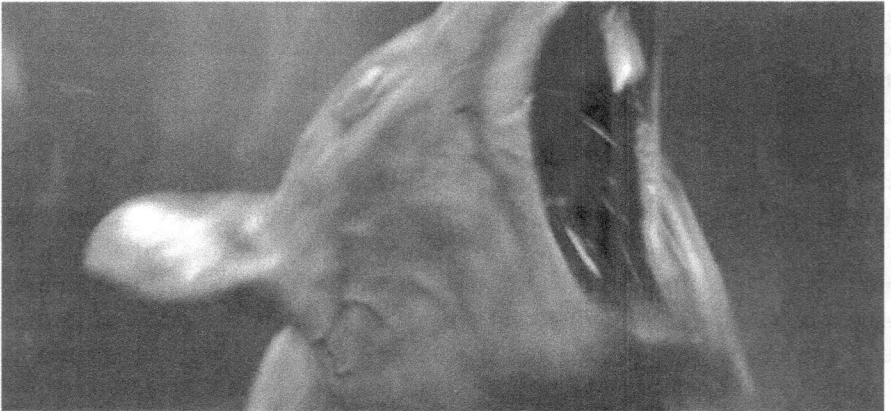

Fig. 2: Rat smashing into the glass wall of its cage

In rhetorical terms, the relation between these two images is a metaphor: the colors, the lights, and the movements are similar, the only difference is that the enraged Darkseeker is substituted by an enraged rat. In this case, the zombification is not determined by a scientific determination of the Darkseeker's objective qualities but rather by the metaphorical exchange of qualities between the image of the Darkseeker and the image of the rat. Neville's hard science is based on a network of soft analogies. In Deleuzian jargon, one would talk about the "becoming-rat" of the Darkseeker, his *devenir-rat*, but then one would probably forget that this transformation is a deprivation rather than an emancipation: the Darkseeker is not free but Neville is free to execute his rat-like test-animals with impunity.

My second example is an apostrophe. In a deserted DVD shop, Neville keeps returning to a pretty female mannequin. After his dog dies, he finally approaches the mannequin, turns his head toward her, starts to cry and talks to her: "I ... I promised a friend I would say hello to you today ... Please say hello to me." In Greek, *apostrephein* means turning away: in an apostrophe, the speaker turns away from his human addressee and directs his speech towards an inanimate object or an abstract concept. In this case, Neville addresses a doll as if it was a human being with which he could enter into some kind of relation. The idea of turning implied in the original Greek meaning of *apostrephein* is dramatized when Neville, in a slow and stylized movement, turns his head toward the mannequin. Talking in apostrophes is basically saying "Please say hello to me" to an inanimate world.

Neville's turning of his head is, in fact, the very turning point of the movie since the apostrophic mode of address creates a real human being. It seems as if the apostrophe has a performative power that can turn a non-human object into a human. The Charlton Heston adaption of Matheson's novel, *The Omega Man* from 1971, refers a little crudely to the Greek legend of Pygmalion and Galatea: as Neville talks to one of the mannequins, she actually is transformed into a living woman. In Lawrence's *I Am Legend*, the performative power of the apostrophe sets in with a delay: Neville leaves the DVD shop and drives down to the docks to lance a suicide attack on the Darkseekers – only to be saved by a real woman, the immune Anna.

To sum up, Neville's interpretation of the infected is informed by a social imaginary that, in Paul de Man's terms, can be defined as a "tropological system" (de Man 1984, 254). The Darkseekers become visible through a dense network of tropes and figures that orchestrate their ontological and legal status.

4 What Are They?

As we have seen, Neville's sovereign decision about the Darkseekers' non-human status is based on a preceding interpretation informed not only by their objective features but also by the network of tropes and figures that frames his ethical responsiveness toward them. Methodologically, this means that we have to distinguish between two kinds of images in the movie. On the one hand, the Darkseekers are fictional figures that move around in the reality of the movie and to which we, as interpreters, can choose to attribute different kinds of symbolic meaning. On the other hand, the rat-metaphor and the mannequin-apostrophe are speech-figures that are not present as objects in the fictional reality but are, instead, applied by Neville to interpret the reality around him (we saw the same at two levels in the first scene of the movie where the very fast car was used as a figure of speech by Dr Krippin only to reappear as a real car in the reality of the movie).

Contemporary research in zombie apocalypses tends to focus on the first kind of images, asking about the symbolic meaning of the zombie. The answer has been that we must interpret the zombie as an image of the hypnotized modern consumer (Shaviro 1993, 92), of the terrorist (Bishop 2010, 29), of the victims of biopower (Canavan 2011, 175), and so on. In this article, however, I have explored the second kind of image. Instead of proposing another interpretation of the zombie figure, I have been analyzing the tropes and figures through which the fictional characters themselves interpret the zombie. If we wish to understand the zombification of social space in *I Am Legend*, the important thing is not what the Darkseekers mean but how they become meaningful to Neville.

Barbara, one of the first zombie victims in film history, cries out "What are they?" in the beginning of Romero's *Night of the Living Dead*. Since then, modern zombie movies include an on-going zombie interpretation which, however, never seems to supply a satisfying answer to Barbara's question. In Colson Whitehead's abovementioned zombie novel the fictional characters' incessant interpretation of zombies is even a dominating theme. *Zone One* is a story about a group of *sweepers*, a kind of volunteer who cleans any remaining zombies out of downtown Manhattan. With ethnographical precision, Whitehead analyses how these *sweepers* interpret the zombies, for instance by giving them imaginative names (a *skel*, a *dead*, a *hostile*, an *unfortunate*, etc.) or by playing callous Abu Ghraib-style games with the killed zombies.

In one scene in the novel, two groups of *sweepers* have just met up with each other on Fulton Avenue in the financial district as a *skel* suddenly comes staggering toward them:

> The skel wore a morose and deeply stained pinstripe suit, with a solid crimson tie and dark brown tasseled loafers. A casualty, Mark Spitz thought. It was no longer a skel, but a version of something that predated the anguishes. Now it was one of those laid-off or ruined businessmen who pretend to go to the office for the family's sake, spending all day on a park bench with missing slats to feed the pigeons bagel bits, his briefcase full of empty potato-chip bags and flyers for massage parlors. The city had long carried its own

plague. Its infection had converted this creature into a member of its bygone loser cadre, into another one of the broke and the deluded, the mis-fitting, the inveterate unlucky [...] This creature before them was the man on the bus no one sat next to, the haggard mystic screeching verdicts on the crowded subway car, the thing the new arrivals swore they'd never become but of course some of them did. It was a matter of percentages (Whitehead 2012, 121).

In this description, it is easy to pin down the symbolic meaning of the zombie: the zombified businessman looks like "a casualty", one of those who ran out of luck in modern capitalism where success is a matter of percentages. The link between zombie and loser is based on the novel's underlying metaphorical analogy between the zombie disaster and modern disaster capitalism. It is important to realize, however, that it is the protagonist that sees the zombie in this way: "A casualty, *Mark Spitz thought*." The description is not about the zombie itself, it is, rather, about the way the *sweepers* perceive the zombie: "This creature *before them* ..." Rather than just using a given social imaginary, the novel is analyzing the images through which the fictional characters interpret their post-apocalyptic world. Before the disaster, the socially shared images of social life were used to draw a small-meshed net of distinctions dividing different social groups from each other; after "Last Night", only one line is drawn, namely the border between humans and *skels*. In the terminology of Jacques Rancière, the distribution of the sensible has turned into a crude dichotomy between Us and Them. This simplification of the social imaginary is analyzed when one of Mark Spitz's *sweeper* colleagues gets bitten by a *skel*:

> He found it unlikely that Gary was not in ownership of a master list of racial, gender, and religious stereotypes, cross-indexed with corresponding punch lines as well as meta-textual dissection of those punch lines, but he did not press his friend. Chalk it up to morphine. There was a single Us now, reviling a single Them (Whitehead 2012, 231).

5 Could They be Evolving?

According to Judith Butler, the practice of critique is the practice of framing the frames. Instead of just seeing the world through the frame, we need to learn to see the frame that blinds us, "exposing the orchestrating designs of the authority who sought to control the frame" (Butler 2009, 12). As we have seen, *I Am Legend* functions as an expository device that lays bare Neville's pre-conscious framing of the Darkseekers as non-human. But does the movie in fact criticize the tropological system that guides Neville's interpretation of the Darkseekers or does it rather confirm it? Is his social imaginary put to test or just put to work?

It is important to note that Neville's picture of the Darkseekers is indeed called into question as the story evolves. Still at the beginning of the movie, Neville sets a snare trap to capture a Darkseeker in order to test a promising serum against the plague. He lures a female Darkseeker into the trap with a little blood just as he would do if he wanted to catch a beaver or a raccoon. Apparently, the rat-metaphor not only determines the way Neville sees the Darkseekers but also the way he acts towards them. The next day, however, Neville is puzzled as he finds one of the mannequins from the DVD shop out in the street in front of Grand Central Terminal. As he approaches this second mannequin, he is caught in a snare trap and passes out. If the Darkseekers have managed to copy his snare trap, the de-evolution of their rational capacities cannot be complete. In one of the deleted scenes, which can be found on YouTube, Neville tries to convince Anna that the Darkseekers could not have set the trap: "That's my snare, these are my materials. The infected didn't do this, they can't." According to Neville's interpretation of the Darkseekers, "they have no higher brain function, they don't plan." Later, however, in another deleted scene, he starts to vacillate: "Could they be evolving?"

It is remarkable that these two scenes were cut out. In the official version of the movie the story of the Darkseekers' evolution, and of Neville's growing awareness of it, is repressed, which turns the scene with the Darkseekers' snare into a meaningless cul-de-sac. The movie simultaneously tells a story about evolution and a story about the eradication of the Darkseekers and this conflict surfaces in the movie's two contrasting endings. In the original ending, now available as extra material on the Two-Disc Special Edition of the movie, Neville, Anna and the boy Ethan retreat to the basement laboratory chased by a horde of Darkseekers that have broken into the house. The three humans, together with the female Darkseeker that Neville caught for his experiments, are still protected behind a reinforced Plexiglas wall when they discover that the female Darkseeker has started to turn into a human being. Apparently, Neville has finally found a serum against the plague. However, time is short since the leader of the Darkseekers is smashing into the protecting Plexiglas wall that begins breaking up.

Fig. 3: Darkseeker attacking plexiglass wall

As the leader of the Darkseekers draws a butterfly-shaped smear on the Plexiglas, Neville suddenly realizes that the female Darkseeker caught on the human side of the Plexiglas has a butterfly tattoo. In other words, it seems as if the leader tries to communicate that he wants the female Darkseeker back, presumably his wife or his daughter. The last time Neville saw his own wife and daughter, the daughter drew a butterfly on the window of the helicopter before it crashed. Thus, the butterfly, a classical symbol of the soul, is a token of the matrimonial or fatherly love shared by Neville and the Darkseeker. Neville puts down the gun, opens the door, and in a kind of wordless contract he gives back the female to the male Darkseeker and is, in recompensation, spared by the horde of Darkseekers who peacefully leave the building. In the terms of political philosophy, this is a social contract turning the lawless state of nature into political society.

In the original ending, both metaphor and apostrophe reappear but in transformed versions. When the male Darkseeker smashes into the Plexiglas wall, he looks very much like the first Darkseeker smashing into the windshield and the rat smashing into the glass wall of its cage. Hence, the picture continues the zombifying metaphor from the first half of the movie. The important difference, however, is that the glass wall is gradually breaking down. What we see in the original ending is a deconstruction, in a very literal sense of the word, of the rhetorical border between humans and non-humans.

While the leader attacks the Plexiglas wall, Neville turns his head towards the Darkseekers outside the wall and talks to them: "I can save you." With his newfound serum, an efficient biopolitical tool, he can provide health and humanity. In rhetorical terms, this is another example of an apostrophe, only this time Neville does not turn his head toward a lifeless mannequin or a speechless dog but, for the first time in the movie, toward the Darkseekers. It seems as if this mode of address is sufficient to turn the Darkseekers into human beings.

If you replace the very bad man with a cop "the picture changes," Dr Krippin explained in the first scene of the movie. In the original ending, the metaphor and the apostrophe are transformed and this changes the picture with the Darkseekers replacing Neville the cop with Neville the very bad man. After the wordless contract between Neville and the Darkseekers, the camera turns toward a wall full of Polaroids of Darkseekers that were used as test animals in Neville's scientific experiments. Our knowledge of the matrimonial or fatherly love of the leader turns these Polaroids into a memorial wall with pictures of crime victims. In the original ending Neville's last words are: "I am sorry." In the new normal after the plague, we understand, the Darkseekers are rights-bearing human beings and not just rightless test animals.

To be sure, the wordless passion between male and female Darkseeker, accompanied by a romantic piano *adagio*, is kitsch. Suddenly, the Darkseekers obtain an ideal humanness defined by traditional family values. A similar de-zombification is found in Neill Blomkamp's *District 9* (2009) in which the non-human aliens, known by the derogatory metaphor *prawns*, suddenly prove to be even more family-minded than the humans. Driving out of New York City towards a survivors' stronghold in Vermont, Neville, Anna and the boy pass an idyllic family of lions with two cute cubs playing in the street. It seems as if the expansion in the scope of the human is balanced by a narrowing of the substance of the human. However, the original ending still revises and retracts the zombifying rhetoric with which the movie starts out. The picture changes.

In the official version, however, the butterfly and the wordless contract are replaced by a hand grenade. Here, the butterfly tattoo, the token of humanness, has moved from the female Darkseeker to Anna, who was a human being from the very start. While the leader attacks the Plexiglas wall, Neville gives a vial of cure-containing blood to Anna so that she can save humanity and he then just grabs a hand grenade and blows himself up along with the Darkseekers. After Neville's dead, Anna and the boy find food, shelter and protection in the survivors' gated community. While Anna is handing over the serum to one of the survivors we hear her voice in a voice-over:

> In 2009, a deadly virus burned through our civilization, pushing humankind to the edge of extinction. Dr Robert Neville dedicated his life to the discovery of a cure and the restoration of humanity. On September 9th, 2012, at approximately 8:49 P.M., he discovered that

cure. And at 8:52, he gave his life to defend it. We are his legacy. This is his legend. Light up the darkness.

Anna's voice-over pays homage to the sovereign violence used by Neville. At the end of the day, he was not a very bad man but a hero cop who sacrificed his life in the struggle to restore humanity. In the original ending, Anna has a parallel voice-over that ends with the words "Keep listening. You are not alone." In the official version, Anna's summons to listen is replaced by a summons to lighten up the darkness. Apparently, there is no need to be ethically responsive to the Darkseekers. In the terminology of the state of emergency "Light up the darkness" means: use justified violence in order to subject heterogeneous social space to the regulations of law. In this way, Neville keeps the promise he makes to his daughter in one of the flashbacks of the movie: "Daddy's gonna make the monsters go away." In a Manichean logic of light and darkness, it is easy to draw the line between the "We", who are Neville's legacy and the rest, who are not included in the victorious "We." In a situation in which a hand grenade is the only weapon at hand, "Light up the darkness" simply means: blow up the Darkseekers.

During Anna's voice-over, the camera lifts to a birds-eye perspective on the survivors' classic American village. In the same movement, the perspective leaves behind the rhetorical network that governed Neville's interpretation of the Darkseekers. In her homage to sovereign power, Anna in the official version speaks in a demonstratively non-figurative language of facts and figures, titles and clock-times: Dr Robert Neville discovered the cure at approximately 8:49 p.m. and gave his life at 8:52. This is not a moment of truth in which the objective facts finally penetrate the "veil of ignorance" created by the social imaginary. Instead, it is a moment of ideology. After the tropes and figures of the social imaginary have created the ontopolitical picture of the Darkseekers, the ladder is kicked away. The inherent violence to Neville's interpretation of the infected is hidden from sight. In this way, a product of rhetoric is transformed into a fact of nature: an unbridgeable and unchangeable chasm between zombies and citizens.

References

Agamben, Giorgio. *Homo sacer: Sovereign Power and Bare Life*. Stanford, CA: Stanford University Press, 1998.
Arendt, Hannah. *The Origins of Totalitarianism*. San Diego: Harcourt Brace Jovanovich, 1979.

Bishop, Kyle William. *American Zombie Gothic: The Rise and Fall (and Rise) of the Walking Dead in Popular Culture*. Jefferson, NC, and London: McFarland, 2010.

Brooks, Max. *The Zombie Survival Guide: Complete Protection from the Living Dead*. New York: Three Rivers Press, 2003.

Butler, Judith. *Precarious Life: The Powers of Mourning and Violence*. London and New York: Verso, 2004.

Butler, Judith. *Frames of War: When is life Grievable?* London: Verso, 2009.

Canavan, Gerry. "Fighting a war you've already lost: Zombies and Zombis in Firefly and Dollhouse." *Science Fiction Film and Television* 4 (2011): 173–204.

Connolly, William. *The Ethos of Pluralization*. Minneapolis: University of Minnesota Press, 1995.

Connolly, William. *Pluralism*. Durham: Duke University Press, 2005.

Drezner, Daniel W. *Theories of International Politics and Zombies*. Princeton, NJ: Princeton University Press, 2011.

Feldman, Leonard C. "The Banality of Emergency: On the Time and Space of 'Political Necessity'." In *Sovereignty, Emergency, Legality*, ed. Austin Sarat, 136–164. New York: Cambridge University Press, 2010.

Fraser, Nancy. *Scales of Justice: Reimagining Political Space in a Globalizing World*. Cambridge: Polity, 2008.

Holm, Isak Winkel. "Earthquake in Haiti: Kleist and The Birth of Modern Disaster Discourse." *New German Critique* 39, no. 1 (2012): 49–66.

Honig, Bonnie. *Emergency Politics: Paradox, Law, Democracy*. Princeton: Princeton University Press, 2009.

Honneth, Axel. *Das Recht der Freiheit: Grundriss einer demokratischen Sittlichkeit*. Berlin: Suhrkamp, 2011.

Kay, Glenn. *Zombie Movies: The Ultimate Guide*. Chicago: Chicago Review Press, 2008.

Lillemose, Jacob, and Karsten Wind Meyhoff. "When there is no more room in hell, the dead will walk the earth: George A. Romeros billeder af den amerikanske undtagelsestilstand." *Kritik* 202 (2011): 50–66.

de Man, Paul. *The Rhetoric of Romanticism*. New York: Columbia University Press, 1984.

Matheson, Richard. *I Am Legend*. London: Millennium, 1999.

Rawls, John. *A Theory of Justice*. Cambridge, MA: Harvard University Press, 1999.

Schmitt, Carl. *Politische Theologie: Vier Kapitel zur Lehre von der Souveränität*. Munich: Duncker & Humblot, 1934.

Schmitt, Carl. *Political Theology: Four Chapters on the Concept of Sovereignty*. Cambridge, MA: MIT Press, 1985.

Shaviro, Steven. *The Cinematic Body*. Minneapolis: University of Minnesota Press, 1993.

Taylor, Charles. "The Nature and Scope of Distributive Justice," in: *Philosophy and the Human Sciences*. Cambridge: Cambridge University Press, 1985.

Taylor, Charles. *Modern Social Imaginaries*. Durham, NC: Duke University Press, 2004.

Walzer, Michael. *Spheres of Justice: A Defense of Pluralism and Equality*. New York: Basic Books, 1983.

Whitehead, Colson. *Zone One*. London: Vintage, 2012.

Inês Espada Vieira

How Do We Measure Disaster? How Do We Ensure Security?

The 2011 Lorca Earthquake in the Media

1 The 2011 Lorca Earthquake

During 2010, the world experienced two major earthquakes with dreadful consequences: Haiti (12 January) and Chile (27 February). In early 2011, another major earthquake hit Japan (11 March) leaving the country to cope – again – with unthinkable destruction and suffering with the historically sensitive nuclear question thrown into the equation.

Considered in this wider context, the earthquakes that took place in Lorca, Murcia, Spain, on 11 May 2011, were not especially significant events: an earthquake of a moderate 4.5 magnitude preceded another of 5.1 Richter magnitude. The epicenter was however located at a very shallow depth, which made its impact stronger. Nine persons lost their lives. The Murcia Region is familiarized with seismic activity and, in a broader sense, Spain has experienced disaster all too well, both natural and manmade.[1] In fact, we could say, as does David Alexander in referring to the Versilian floods of 1996 that, in the global context, this was a "truly unexceptional event" (Alexander 2000, 2).[2]

One of our starting questions was to try to find out how do we measure disaster. As stated by the U.S. Geological Survey (USGS) and its Earthquake Hazards Program, the 2011 Lorca Earthquake was one of type VI intensity, with "strong" shaking, and "light" damage.[3] For the Spanish National Geographical Institute and its Seismic National Net, the earthquakes in Lorca returned an

---

1 The *Prestige* oil spill in 2002 was one of the major ecological catastrophes in contemporary Europe and in Spain it has become a leading object of academic research as seen by the number of books and articles published on the subject. The most important and complete Spanish bibliographical web portal DIALNET offers an extensive list of these works (http://dialnet.uni rioja.es/).

2 Between the years 1945 and 1986, there were 2.34 million deaths in disasters, with an average of 56,000 deaths and 30 disasters per year (Glikman 1992, *apud* Torrence and Grattan 2005, 2).

3 http://earthquake.usgs.gov/earthquakes/dyfi/events/us/c0003c5s/us/index.html (accessed 26 October 2011).

intensity of VII ("damaging") according to the European Macroseismic Scale EMS-98.[4]

There are, among other tools, geophysical measurement instruments. However, our point of view is not of a technical nature. Bearing that in mind, we analyze the specific case of the 2011 Lorca Earthquakes[5] and the way the Spanish press contributed to measuring the event. After defining its "size",[6] we observe different strategies from the different participant agents in the aftermath of the catastrophe whose symbolic actions were meant to ensure (a sense of) safety.

2 How Do We Measure Disaster? The Size of a Disaster

Although the difference between natural, technological and social hazards[7] is still an effective and operative factor for scientific analysts and civil protection authorities in different countries, we also know that there is no such thing as a "pure" natural disaster as human and material loss is often a consequence of bad and careless decisions taken on issues such as socio-economic activities and soil use in general. In Portugal, we have a sad example of this issue with the 2010 Madeira floods, leaving 42 dead and 100 injured, and not only a consequence of heavy rain but also of unplanned urban development and overly eager construction very near or on watercourses.

For the victims or/and, as is so often used in Spanish, for those affected (*afectados*), every disaster is unique (Alexander 2000, 2). In a diachronic and geographical perspective, it becomes obvious that each event, as with each day in every place around the globe, happens only once. Catastrophic events are

---

4 The European Macroseismic Scale was defined in 1998 by the European Seismological Commission and measures not the magnitude – the energy release – of the seismic event but rather its consequences, the way it affects people and buildings. This represents a sort of "manual" for earthquake assessment in which geophysicists worked together with engineers to set a reference for the measurement of intensity (see Grünthal 1998).

5 In order to represent Spanish usage, we refer to 11 May Lorca Earthquakes in the plural. The earth shook various times, however, it were the two stronger quakes that proved damaging.

6 "[...] to some extent the size of disaster may be independent of the magnitude of the geophysical event" (Alexander 2000, 8).

7 See, for instance, the Vademecum of Civil Protection of the European Commission, which distinguishes between two types of disaster: "Natural and man-made disasters" (http://ec. europa.eu/echo/civil_protection/civil/vademecum/index.html (accessed 10 January 2012).

always singular and unique for those who live them but also, apparently, for those who report them. When reporting a catastrophe, the press, for instance, try to convey an image of singularity by choosing personal stories and individual testimony of several situations. Nevertheless, we are all aware that this happens only at a superficial level of reading: the singular events reported in the news are actually not that singular, they are exemplar, building the narrative of the group affected by that particular event and following already established models. Hence, those stories represent only one, and always the same, history of the catastrophe which is repeatedly reported during its immediate aftermath: a certain scheme of "stereotypical reporting" which was only exacerbated by the development of "rolling" or "24/7" news (Benthall 2010, xiii). Jonathan Benthall argues that there is a stable narrative structure in the way journalists tell stories of disaster relief which the author compares with the folk tale structure presented by Vladimir Propp in his *Morphology of the Folk Tale* (1928): "[...] a donor sends the hero, armed with magical powers, to combat evil or famine, till he returns home from his dangerous mission to receive congratulations from a princess" (Benthall 2010, xvii). Benthall refers specifically to international emergency aid teams working in distressed areas but we can identify (if not the same obvious structure) at least the same primacy of the hero (distinct heroes, actually) in the reporting of disaster.

Suffering is one of the favorite representations of all times in western culture ever since the Greek tragedy, which some authors still find as the model for the contemporary media discursive structure for reporting a catastrophic event. García Gómez argues that the media narratives of catastrophe obey a model that somehow follows the structure of classical tragedy: "From the point of view of the professional communicator who must face these occurrences, the catastrophe is seen as a tragedy (in the technical sense of the term): a type of narrative whose *pathos* stems from the dialectic between *broken order/restored order*" (García Gómez 2006, 126). I am not saying that it is a genre, or that disasters follow a certain exact model, but rather that reporters tend to use "tragicity" as a model for their "stereotypical" discourse, thus simplifying the information about a reality which is never that simple. Using this dramatic vocabulary, R. Kapuscinski affirms that we are actually facing a "new dramaturgy", one "that has no final act, no ending". The reporters and the news turn away from the site and from the subject as if there had been no event (Kapuscinski 2000, 117). Our attention is drawn away from the scene making it seem that the disaster is over but it might not yet be and even extending into the near future.

In *Distant Suffering: Morality, Media and Politics* (1999),[8] Luc Boltanski writes about the politics of pity and reminds us "that one of the main motivations of fiction is the staging of suffering and that the spectacle of suffering has been seen as a cause of the spectator's pleasure" (Boltanski 1999, 21). However, this suffering may not necessarily be cruelly enjoyable to the public but rather something seen with sorrow, that seeks a reaction of pity and solidarity towards those who are in pain. If we see news reporting as a narrative based on "real" facts, with the same structure as any fictional narrative, we can understand the power of displaying pain and suffering through the pictures published in the day-after newspapers, where we can observe the distinct levels to the process of making the dignity of people vulnerable (Oyanedel and Alarcón 2010, 117).

3 The Front Pages of 12 May 2011. The Day after the Earthquakes

I have chosen twelve titles both from the national and the regional press. Apart from four of them, all these newspapers made recourse to two photos of the same scene (with slight differences in that they belong to the same sequence of images), taken for EFE (the Spanish news agency) by reporter Israel Sánchez. In both photos, we see all the basic features of a catastrophe:
- the scenario – destruction – wreckage;
- the main theme – death – the corpse of a victim lying on the ground;
- the characters;
- pain – the woman crying supported by other two;
- aid – two women (a police officer and a citizen).

There are other people in these photos. We see another police officer talking on the walkie-talkie; some civilians passing by (a man in a pink shirt and an old woman in black). In one of the pictures, we also see a civilian (and a dog!) leant over the corpse.[9]

---

8 Originally in French: Luc Boltanski. *La souffrance a distance: morale humanitaire, médias et politique*. Paris: Éditions Métailié, 1993. Boltanski argues that the spectator should react to the suffering s/he sees in the news through three rhetorical topics: denunciation, sentiment and aesthetics.
9 The front page of *La Verdad*, with one of these pictures, was awarded best daily newspaper front page at the 8th edition of the ÑH Awards for the Best News Design Spain & Portugal (Society for News Design) (http://www.snd-e.com/en/nh/premios-nh (accessed 12 January 2012).

As for the titles and not surprisingly, we find the words "tragedy" (*La Verdad*), "mortal" and "death" (*ABC, La Razón, El Periódico, Público, El País, Levante*), "terror" and "terrified" (*Las Provincias, El País*), "panic" (*La Gaceta*), "horror" (*La Opinión*). *La Vanguardia* chose to make the earth the subject of the action and Lorca the object, in our view punctuating the leading role of nature in the hazard: "Earth shakes Lorca".

The online edition of *El Mundo* also used the photo by Israel Sánchez. However, in the paper edition the day after the earthquakes that photo appeared with a cut that excluded the corpse in the left background. The director, Pedro J. Ramírez, published the two versions of the front page and explained by Twitter[10] that the change was in answer to tweets asking the paper not to print the image of a dead person.

As for the other photographs, *Levante* and *La Opinión* (both regional papers) chose a picture of a dead body not completely covered by a metal blanket and lying in the street full of building debris. The strong shining of the "gold" makes a sharper contrast with the blood we can see. The picture is morbidly colorful, stressed by the yellow and blue of what seems to be a helmet and the pair of sneakers. The original photo was larger and contained another person. Emphasizing the solitude of the moment, this cut brought death into the foreground as the protagonist of the hazard.

Taking place precisely two months after the Japan Earthquake, only two newspapers noticed this coincidence in the dates. *La Verdad* drew attention to Japan in its "fake" front page,[11] making an indirect analogy (the form of presentation makes us believe that it had already been prepared). In turn, using one of the photos we mentioned, *La Gaceta* (from the Christian conservative media group Intereconomía) entitled the news by making the correspondence with Japan: "During five hours Murcia lived the panic of Japan".

The daily paper *ABC*, printed in Madrid and one of the oldest newspapers in Spain, chose a picture by graphic reporter Nacho García displaying a medical doctor from the Spanish emergency services (María José Carrillo) with a wound-

---

10 On 12 May, Pedro J. Ramírez tweeted: "Bdías. Todos con Lorca. La portada d EM. Esta era la prevista http://bit.ly/mDXI5j. Y esta la q salió d vuestros tuits http://bit.ly/iMjJgR." http://twitter.com/#%21/pedroj_ramirez/status/68558765997297664 (accessed 31 October 2011).

11 The printed edition of the newspaper used the main photograph covering the total extent of the newspaper, both its front and last page. However, online, *La Verdad* used a different version of that day's edition with a "normal" front page: paper title, publicity, main theme with the picture as well as other stories.

ed child (Sergio, 3 years old) in her arms.[12] In the background, we can see many other rescuers (professional or civilians) and the ruins of buildings. In this picture, the doctor is the hero. The heroes are always one of two species: the rescue personnel (nurses, doctors, firepersons, etc.) and the "anonymous" citizen that exceeds his or her common life role and does something considered extraordinary and enlarged by the extraordinarily hazardous event.

In the days after the event, the news divided between data about the people deceased, injured and homeless, and the stories of those participants: the pregnant women who died, those in the hospital, all the people spending the night in the open air …

We should always bear in mind that pictures are not objective, that we need to decode them, images do not just communicate with us, they have to be interpreted and, as the doctor that specializes in ultrasounds or scanners, the more we observe and study, the better the analyst we become. The pictures we have been commenting on do not show any particularity of the Lorca Earthquakes, they could belong to any other disaster as they always tell the same narrative: the wounded, the death, the saviors, the children, the suffering, the impotence of mankind before nature. This is a form of taking the disaster out of context, focusing on the consequences (death and suffering) and not on what really happened (the earthquakes in Lorca). In a way, it is the non singular that prevails (death and suffering), although, as we said, the media seek to tell the story of a singular event (the earthquakes in Lorca).

So, let us return to the beginning question: "How do we measure disaster"? The media set out their own measurements, apart from the geophysical data. The number of wounded, the number of destroyed structures, the amount of the cost of rebuilding and, most importantly, the number of deceased. The victims are seen as "the critical ingredient of a *disaster*" (Torrence and Grattan 2005, 5). Is nine dead a "real" catastrophe? For how long is it a catastrophe? In a world of so many hazards, in a year of so many changes, is nine enough to keep a subject in the news?

As some newspapers stated, the 11 May Lorca Earthquakes were the deadliest in Spain since 1956 but the country and the globe offered the temptation of too many "good" news stories during the days that followed: the Strauss-Kahn *affaire* (14 May), the *Indignados* (15 May), the regional elections (20 May) … When something out of the ordinary happens, such as the Lorca Earthquakes, the public faces an "information flood" that provokes a certain "discursive satu-

---

12 This story was one of the most productive in the aftermath of the disaster, explored in different newspapers and online in all its details – the dead mother, the two brothers …

ration" (Bernardo and Pellisser 2010, 109–110) due to the amount and the velocity of the data told. However, after the initial period of shock and mourning, the novelty and the agenda setting criteria are stronger than the initially "breaking news". As always, gradually, over time, the media highlights change to other subjects until (almost) total silence. Bernardo and Pellisser state that the media leave the site where the disaster took place when the iconographic revenue is worn-out, "when the drama transfers from the outside (the street) into the inside of homes" (Bernardo and Pellisser 2010, 111).

Exactly six months after the earth shook in Lorca, only one newspaper – *La Verdad*[13] – dedicated its front page to the anniversary of the disaster.

4 How Can Safety be Ensured?

For quite some time now and right at the moment of writing these lines, Spain is witnessing minute by minute coverage of the uncertain developments surrounding volcanic activity on El Hierro in the Canary Islands. Everyone is on alert – politicians, civil protection, military forces, scientists, people ... – and the media makes daily headlines and on-line live feeds on the event: personal stories, scientific data, crisis management plans, volcanic history, safety recommendations ... Despite – or because of – such an important amount of information, one of the general feelings is the sense of immense human vulnerability in face of the phenomenon. Before nature, we stand always vulnerable and ignorant. And, since the secularization of catastrophe with the 1755 Lisbon Earthquake, we stand alone without any God to justify the disaster, just men and women enduring it.

States whether individually or in groups (such as the European EUR-OPA, the UN or NATO) establish contingency plans to act in the face of distinct natural and technological hazards. In Spain, for instance, Civil Protection (answering to the Interior Ministry) has different action plans for each region, according to the results of risk assessments undertaken to define the needs and probability of certain disasters taking place in those areas of the country. As is logical, there are no sea rescue plans for interior regions such as Madrid or Castilla-León, and no plans for dealing with a heat wave in the north, in Galicia or the Asturias. Nowadays, European countries are ready for almost every natural hazard possible. And, when a state experiences deficiencies concerning some

---

13 *La Verdad* is the most important paper in the Murcia Region. With an average of 259,000 readers per day, it is the 14th most read newspaper in Spain.

particular aspect of disaster preparedness, there are bilateral and multilateral agreements to help mitigate the consequences of catastrophe. Thus, Portugal receives help and support from Spain or France to combat wildfires especially during summer ... every summer.[14]

Many of the measures taken to try to mitigate the consequences of an earthquake (I use the example because it is my subject today) are often not very popular, may be expensive and are mostly invisible to the public: anti-seismic construction, historical buildings conservation, building maintenance, sustainable urban planning, etc. Undertaken before the disaster, these are the most effective measures[15] and involving a great deal of respect for nature and prospective thinking.

As a consequence of seismic activity in the region in 1999 (the main earthquake on 2 February; 4.8 M_w), 2002 (the main earthquake on 6 August; 4.6 M_w) and 2005 (the main earthquake on 29 January, 4.6 M_w),[16] a special project about seismic hazards was undertaken for the Spanish region of Murcia.[17]

The risk was/is known. However, being prepared to face disaster does not mean it can be avoided even if predicted.

No prevention actions or plans are ever enough to make people feel safe and to secure property. Therefore, in the aftermath of a catastrophe, the general goal is to save lives and to limit damage; ensuring safety is a concern of all the institutional, civil, military and humanitarian forces involved in such emergency actions.

At some point, all catastrophe emergency response plans may seem as though the authorities are overreacting if the disaster finally does not actually take place or happens with less strength and impact. The contrary also hap-

---

14 See the webpage of Portuguese National Authority for Civil Protection http://www.prociv. pt/RI2/Pages/RelacoesInternacionais.aspx (accessed 4 January 2012).

15 For the European Environment Agency, one of the reasons of the "relatively low human death toll" in earthquakes in Europe in the 2003–2009 period stems from "the vulnerability of the buildings and their degree of compliance with building codes [...]. Not surprisingly therefore, the effect on both humans and buildings was much more significant in areas where building codes have not been properly applied, leaving people in highly vulnerable buildings, despite being exposed to seismic risk" (EEA, 2010: 98–99).

16 Source: http://www.proteccioncivil-andalucia.org/Informes/SismoLorca29-02-2005.htm (accessed 10 January 2012).

17 The aim of the RISMUR (Seismic Risk in the Murcia Region) Project involves drafting seismic hazard maps, based on which seismic risk maps of the Murcia Region will be developed. The ultimate objective of the project is to provide recommendations for the development of a mitigation plan to reduce seismic risk in the Murcia Region. The RISMUR Project was published in January 2006 by ING (the National Geographical Institute).

pens: reality might always be uglier than we thought it would be and humans have often underestimated the power of nature. We had a tremendous example of that in Fukushima last year. I would say that societies should act like the wise virgins of the Bible who did not know when their spouses would return but conserved enough oil to endure a long wait: "Therefore keep watch, because you do not know the day or the hour" (Mathew 25:13).

Unlike manmade disasters, natural disasters are largely presented by the media as uncontrollable, random events, impossible to predict, avoid or defeat. Therefore, many journalistic discourses tend to a "naturalization" of catastrophe and avoiding the discourse of "communicative responsibility" that requires a more profound explanation of the events and instead slipping into immediate and easy spectacular journalism rather than any reflection on the causes, the process and the effects of the disaster (Bernando and Pellisser 2010).

Political, state, civil and military authorities often depend on the press, TV, radio and the Internet to spread messages to the public including practical information as well as updates on various situations. However, the media can also reveal the fragilities of the response strategies. Whenever this happens, problems may arise involving still greater fear or even panic because, as García Gómez states, "The most important function of public communication is not to *inform* about what is happening, but rather to *reestablish confidence in the mechanisms of social organization*" (García Gómez 2006, 130). Through the media, the public must understand that although it might take some time, social organization will go back to "normal" and normalcy will resume.

In modern Western societies, the visibility of the disaster stems from media interest in the event, and, as already seen, is very much determined by the resultant number of casualties. After a certain period, media attention, and consequently society's, draw back from the field and move onto another place, another subject and another (extra)ordinary event. "When the dead are buried the theme no longer has a future" (Bernardo and Pellisser 2010, 112). However, "the future" often returns in the form of an exceptional visit from politicians or celebrities (from music, cinema, sports ...), who might draw the public's attention again to the place of the former events. We see all this constantly and not only with natural disasters but also with other emergency situations in developing countries. These are exceptional visits and they often represent an effort from humanitarian aid agencies or local authorities to revive the subject in the news and, hence, revitalize the reconstruction of buildings and lives.

Other visits follow that usual process in the aftermath of a catastrophe. In the "official support and visits to the place" category, everything happened according to that supposed to occur after the earthquakes in Lorca: the political parties canceled the campaign, the party leaders made statements declaring their solidarity with the affected, the politicians who were abroad announced their early return to Spain, official telephone calls were made, the prime minister and the defense minister visited Lorca as well as other official (and religious) authorities. The presence of so many public personalities in any one locality would be completely out of the ordinary in normal circumstances; however, it is the exceptionality of the event that makes these actions unexceptional. The media reported every visit, action and reaction in detail, fulfilling not only the need for information for the public but also the need to occupy time and space in their news coverage.

The immediate aftermath of disasters such as earthquakes – that, due to the specificity of the phenomenon, is not always perceived as the moment *after* but still the *exact* moment – can be controlled by fear. In Lorca, thousands spent the night in the streets fearing that the earth would shake again. Emergency teams played the most important role in reestablishing normality as soon as possible whilst the presence of public personalities proves to those affected that they have not been forgotten, that they are especial and that it is now safe. In these terms, the most important visit was that of the crown prince and princess, who attended the official funerals of the victims.

Their visit to the affected area is protocol after a disaster and thereby in the context not extraordinary. The presence of the prince and princess stresses the gravity of the situation and at the same time eases the tension because of the immense symbolic nature of the moment. Although everyone knows it will happen, no one can renounce – and no least the media – the trope of royalty bending to salute the people, bringing comfort to the subjects who, ultimately, are the sole justification for the prince's public role.

Commemorative events held on the seventh day after the earthquake represents protocol after a disaster and thereby in the context are not extraordinary. The presence of royalty stresses the continuing gravity of the situation after one week with further events taking place one month after with others on the annual anniversaries. However, the 7th day provides a kind of a special closure – the entire Real Madrid football team, including the coach and the president, visited Lorca under the motto "Everybody with Lorca" (*Todos con Lorca*). The photographs (those printed in the newspapers as well as those published by the online press and on Real Madrid's official webpage) are astonishing because the joy contrasts so sharply with what had just happened.

After the big reception in the City Hall (both inside with the local authorities and outside with the people), the team visited the most affected area by bus to discover the true dimension of the damage but did not actually stop to view it. Thereby, there are no pictures of damage or wreckage. Thus, in a way, this visit also makes the disaster invisible through keeping it out of the ordinary media's attention. The power of these visits is overwhelming even if seemingly little more than a marketing show. One of the stories reported on the day's news was that of the presence of the father of one of the dead victims who went with his children to see the football team and actually thanked them for coming.

5 Conclusion

As participants or *in absentia*, when we find ourselves looking on at disasters, even those that are geographically or chronologically distant from us, we bow to the overwhelming power of nature.

It is a common sense observation that natural hazards are more dangerous in developing countries, usually leaving behind a greater number of victims and material losses. That is due to the level of disaster preparedness in developed countries that, through "prevention, planning and relief" (the motto of the Portuguese civil protection), help mitigate the consequences of disaster.

Thus, it is the presence of the media that in a way defines its size and its impact.

Apart from all the measurement indicators there are to assess the size of a disaster, there is one that holds the attention of the media and the public: the number of casualties. Thus, after the initial shock, the first mitigation actions in the immediate aftermath, the emergency aid, the official visits of political authorities and the funerals of the victims, the subject becomes rather a local interest and the national attention moves onto other events that, for their exceptionality (based on an interpretation of what occurs, when it occurs, how often it occurs and its perceptible consequences),[18] take the lead in the news agenda.

In order to ensure safety after catastrophe, the media play a key role because, apart from all their spectacular discourses, apart from the seduction of dramatizing suffering, from displaying the victimization of the subject, the media report symbolic moments, symbolic actions that can offer relief to those affected and thus a sense of safety which, while not preventing other losses, may prove effective in providing real comfort to those feeling lost.

This is not real safety but a constructed sense of safety, which can also be, as it is so perceived by the people, real safety itself. Ensuring safety to the people should be a task for before catastrophes. That is to say, true safety must

---

18 Bearing in mind that "the fundamental determinants of hazards are location, timing, magnitude and frequency" (Alexander 2000, 7), we considered these characteristics for any exceptional event.

come from disaster preparedness. Disaster preparedness is related not only to all the prevention plans and human and technological means to prevent or mitigate a catastrophe. This also relates to the education of society to expect, receive and react to hazardous events. Moreover, this has to do with a certain culture of disaster that passes through groups and generations and is reflected in many artistic expressions but also in community traditions, in familiar memory … Thus, that preparedness is a task for national authorities and/or international organizations (political, military or civilian) but is also an effort involving society as a whole. Within this scope, the media play a very important role: symbolic moments featured by special actors are conveyed through and by the media (messenger and participant in the aftermath of disaster). "[…] audiences are given a vision of what is happening, that again legitimizes the organization and the shared social values" (García Gómez 2006, 131).

A while after a disaster, the media are supposed to show us that everything in society is again working (slowly or otherwise). The victims may want to forget as was suggested by one of Real Madrid's football players and maybe go back to normal: "It's over. We are safe now."

Are we? Ever?

References

Alexander, David. *Confronting Catastrophe*. Hertfordshire: Terra Publishing, 2000.
Benthall, Jonathan. *Disasters, Relief and the Media*. Wantage: Sean Kingston Publishing, 2010 (1993).
Bernardo, José María, and Nello Pellissier. "La 'naturalización' mediática de las catástrofes. Una aproximación crítica." *Cuadernos de Información* no. 26 (January-June) (2010): 103–114.
Boltanski, Luc. *Distant Suffering: Morality, Media and Politics*. Cambridge: Cambridge University Press, 1999.
European Environment Agency (EEA). *Mapping the Impacts of Natural Hazards and Technological Accidents in Europe. An Overview of the Last Decade*. Luxembourg: Publications Office of the European Union, 2010.
Garcia Gómez, Andrés. "La estructura narrativa de sucesos catastróficos en los medios de comunicación." In Juan de Díos Ruano Gómez (dir.), *I Jornadas sobre gestión de crisis. Más allá de la sociedade del riesgo*, 125–134. Coruña: Universidade da Coruña, Servizio de Publicacións, 2006.
Grünthal, Gottfried (ed.). *European Macroseismic Scale 1998 EMS-98*. Luxembourg: Cahiers du Centre Européen de Géodynamique et de Séismologie, Volume 15, 1998.
Kapuscinski, Ryszard. *Lapidarium IV*. Barcelona: Anagrama, 2000.
Oyanedel, Regina, and Claudia Alarcón. "Una mirada al tratamiento televisivo de la catástrofe". *Cuadernos de Información* no. 26 (January-June) (2010): 115–122.

Torrence, Robin, and John Grattan (eds.). *Natural Disasters and Cultural Change*. New York and
London: Routledge, 2005.

Newspapers of 12 May 2011:

ABC
El Mundo
El País
El Periódico
La Gaceta
La Opinión
La Razón
La Vanguardia
La Verdad
Las Provincias
Levante
Público [on-line edition]

Front pages of 12 May (in alphabetical order):

Terremoto mortal en Murcia

Al menos ocho muertos y un centenar de heridos en Lorca, en el peor seísmo en medio siglo · Los partidos suspenden la campaña

Una médico retira a un bebé herido tras el derrumbe de un edificio en Lorca

Zapatero improvisa una reunión para evitar a Papandreu

▶Cancela un viaje a Oslo con primeros ministros «intervenidos» y cita hoy en La Moncloa a sindicatos y patronal [42]

Dinamarca cierra fronteras y liquida el libre tránsito en la UE [18]

empieza a ahorrar ya
ADSL + llamadas
por 15,95 €/mes

llama gratis al 1472
orange.es

DEPORTES

El empate con el Levante da al Barcelona su tercera Liga consecutiva [70]

ABC

• No hay muerte natural... toda muerte es un accidente (Simone de Beauvoir) •

EL✹MUNDO

JUEVES 12 DE MAYO DE 2011. AÑO XXIII. NÚMERO: 7.811. EDICIÓN NACIONAL. PRECIO: 1,20 €

La juez pide «objetividad» a la Guardia Civil en el 'caso Marta Domínguez' / Pág. 56

La hija de uno de los fallecidos a causa del seísmo es consolada, ayer, por una agente municipal y varios vecinos en la calle Galicia de Lorca. / ÁNGEL SÁNCHEZ / EFE

Los rebeldes derrotan en la sitiada Misrata a las tropas leales a Gadafi

JAVIER ESPINOSA
Misrata (Libia)
Enviado especial

El grito de victoria de los revolucionarios se escuchó ayer en la sitiada Misrata. Los insurgentes lograron hacerse con el aeropuerto y la Academia del Aire, las principales bases militares de Muamar Gadafi, lo que supone un importante golpe para el líder libio, que pierde el control de la tercera ciudad del país. Los habitantes de la ciudad asediada desde hace más de dos meses por las fuerzas del coronel comenzaron ayer a respirar más tranquilas, aunque el conflicto todavía está lejos de acabar   Sigue en **página 30**

Al menos 8 muertos en Lorca tras dos fuertes terremotos

Un seísmo de 4,4 grados sacudió la localidad murciana a las 17 horas / A las 18.47 sufrió otro de 5,2 grados que fue letal y pudo sentirse en Jaén, Sevilla y Madrid

JAVIER ADÁN / Murcia
Corresponsal
La tierra tembló ayer en Lorca. A las 17 horas, hubo un seísmo de 4,4 grados. A las 18,47, otro de 5,2 grados provocó la muerte de al menos ocho personas, una de ellas embarazada

'La monstruosa explosión nos hizo enmudecer. Fue aterrador cuando sentimos que el suelo del instituto se movía'

y otra menor de edad. Durante horas, el pánico se apoderó de esta ciudad murciana de 92.000 habitantes, 10.000 de los cuales fueron desalojados. Una multitud salió despavorida a la calle, tratando de escapar de un suceso que dejó decenas de heridos,

edificios en ruinas y muchos destrozos. Los seísmos se sintieron en Almería, Granada, Jaén, Málaga e, incluso, Sevilla. También tuvieron eco en Castilla-La Mancha y Madrid. El alcalde pidió anoche agua, comida y mantas.   Sigue en **página 20**

Tercera Liga para el Barcelona de Guardiola que flojeó ante el Levante (1-1)

Páginas 50 a 53

Un grupo de simpatizantes de la izquierda 'abertzale' trata de boicotear, ayer, la intervención de Yolanda Barcina en una plaza de Basaburua presidida por una pintada en favor de ETA.

La legalización de Bildu desata otra vez la coacción 'abertzale'

Los radicales sabotean un acto de Barcina (UPN) en un pueblo gobernado por ANV y donde sólo se atreve a presentarse Bildu

MIGUEL M. ARIZTEGI
La izquierda abertzale radical mostró ayer su cara más intolerante y fanática para recibir a la candidata de Unión del Pueblo Navarro (UPN) a la Presidencia de la Comunidad Foral, Yolanda Barcina. Una docena de radicales del municipio de Basaburua, controlado por la ilegalizada ANV

–que se hizo con los siete concejales del Consistorio por la imposibilidad del resto de partidos de

El Gobierno recurrirá ante el TS sólo si hoy aparecen más pruebas
Página 5

encontrar candidatos que se atrevan a desafiarles en las urnas, igual que ahora ocurre con Bildu–, boicotearon la intervención de Barcina en el concejo de Aizaroitz y la recibieron con una pancarta que se servía del eslogan electoral de los regionalistas para ridiculizarlos.   Sigue en **página 4**

El Mundo [published front page]

● No hay muerte natural... todo muerte es un accidente (Simone de Beauvoir) ●

EL●MUNDO

JUEVES 12 DE MAYO DE 2011. AÑO XXII. NÚMERO 7.611. EDICIÓN NACIONAL. PRECIO: 1,20 €.

La juez pide «objetividad» a la Guardia Civil en el 'caso Marta Domínguez' / Pág. 54

La hija de una de las fallecidas a causa del seísmo es consolada, ayer, por una agente municipal y varios vecinos en la calle Galicia de Lorca.

Al menos 8 muertos en Lorca tras dos fuertes terremotos

Un seísmo de 4,4 grados sacudió la localidad murciana a las 17 horas / A las 18.47 sufrió otro de 5,2 grados que fue letal y pudo sentirse en Jaén, Sevilla y Madrid

JAVIER ADÁN / Murcia
Corresponsal
La tierra tembló ayer en Lorca. A las 17 horas, hubo un seísmo de 4,4 grados. A las 18.47, otro de 5,2 grados provocó la muerte de al menos ocho personas, una de ellas cuatalmente.

'La monstruosa explosión nos hizo enmudecer. Fue aterrador cuando sentimos que el suelo del instituto se movía'

y otra menor de edad. Durante horas, el pánico se apoderó de ésta ciudad murciana de 92.000 habitantes. 10.000 de los civdes fueron desalojados. Una multitud salió despavorida a la calle, huyendo de escapa de un sucesso que dejó decenas de heridos.

edificios en ruinas y muchos destruidos. Los seísmos se sintieron en Almería, Granada, Jaén, Málaga e, incluso, Sevilla. También tuvieron eco en Castilla-La Mancha y Madrid. El alcalde pidió anoche agua, comida y mantas.
Sigue en página 20

Los rebeldes derrotan en la sitiada Misrata a las tropas leales a Gadafi

JAVIER ESPINOSA
Misrata (Libia)
Enviado especial

El grito de victoria de los revolucionarios se escuchó ayer en la sitiada Misrata. Los insurgentes lograron hacerse con el aeropuerto y la Academia del Aire. Las principales bases militares de Muamar Gadafi, lo que supone un importante golpe para el líder libio, que pierde el control de la tercera ciudad del país. Los habitantes de la ciudad asediada desde hace más de dos meses por las fuerzas del coronel continúan ayer a vislumbrar más tranquilas, aunque el conflicto todavía está lejos de acabar.
Sigue en página 30

Tercera Liga para el Barcelona de Guardiola que flojeó ante el Levante (1-1)

Páginas 50 a 53

Un grupo de simpatizantes de la izquierda 'abertzale' trata de boicotear, ayer, la intervención de Yolanda Barcina en una plaza de Basaburua presidida por una pintada en favor de ETA.

La legalización de Bildu desata otra vez la coacción 'abertzale'

Los radicales sabotean un acto de Barcina (UPN) en un pueblo gobernado por ANV y donde sólo se atreve a presentarse Bildu

MIGUEL M. ARIZTEGI
La izquierda abertzale radical mostró ayer su cara más intolerante y fanática para recibir a la candidata de Unión del Pueblo Navarro (UPN) a la Presidencia de la Comunidad Foral, Yolanda Barcina. Una docena de radicales del municipio de Basaburua, controlado por la ilegalizada ANV,

«que se hizo con los siete concejales del Consistorio por la imposibilidad del resto de partidos de

El Gobierno recurrirá ante el TS sólo si hoy aparecen más pruebas
Página 5

encontrar candidatos que se atrevían a desafiarles en las urnas, igual que ahora ocurre con Bildu-, boicotearon la intervención de Barcina en el concejo de Arantza y la recibieron con una pancarta que se servía del eslogan electoral de los regionalistas para ridiculizarlo.
Sigue en página 8

El Mundo [initial front page attempt; not published on paper]

EL PAÍS

EL PERIÓDICO GLOBAL EN ESPAÑOL

www.elpais.com

JUEVES · 12 DE MAYO DE 2011 · Año XXXVI · Número 12.476 · EDICIÓN MADRID · Precio: 1,20 euros

El Barça conquista su 21º título de Liga

▸ El punto final. Los azulgrana empatan ante el Levante con un gol de Keita (1-1)
▸ "El Madrid, gran rival". Los barcelonistas, unánimes: "Ha sido muy duro"
▸ Pep y las angustias. Guardiola llegó a plantearse dejar el banquillo   **PÁGINAS 50 A 56**

Una policía y una vecina consuelan a una mujer, en la calle de Galicia en Lorca, donde una víctima yace entre los escombros. / MARCIAL GUILLÉN (EFE)

Dos fuertes terremotos causan ocho muertos en una aterrorizada Lorca

▸ El segundo seísmo provoca múltiples daños en la ciudad murciana
▸ Zapatero y Rajoy acuerdan suspender hoy los actos de campaña

JAVIER RUIZ, Lorca

Dos terremotos sacudieron ayer la ciudad murciana de Lorca (92.000 habitantes) y causaron al menos ocho muertos y 113 heridos, tres de ellos graves, como consecuencia del derrumbe de edificios y cornisas. Los seísmos, de magnitud 4,5 la siete kilómetros del casco urbano al noreste) y 5,1 (próximo al centro de la ciudad), golpearon la localidad entre las 17.05 y las 18.47, según el Instituto Geográfico Nacional, seguidos de más de una veintena de réplicas. La localización del segundo seísmo en el casco urbano, además de su magnitud, fue lo que agravó sus consecuencias.

Un total de 10.000 personas fueron desalojadas por el riesgo de réplicas. Muchas de ellas pasaron la noche en la calle. Lorca ya fue destruida por terremotos en 1674 y 1818. Los líderes del PSOE y del PP acordaron ayer suspender un día la campaña [en señal de duelo]. Rodríguez Zapatero y Rajoy viajarán hoy a Lorca. **PÁGINAS 11 y 12**

"El suelo estaba vivo, se resquebrajaba"

Miles de lorquinos deciden pasar la noche al raso por temor a las réplicas

Salían a la calle desorientados por el temblor. "No sabíamos si ir a un parque o qué hacer", relataba ayer Paloma Sanz, vecina de Lorca, unos minutos después del segundo terremoto. Eliseo López contaba que en el concesionario donde trabajaba "saltó todo el alicatado del suelo, resquebrajándose como si estuviera vivo". En el concesionario hubo un herido, golpeado por un trozo de techo. A Cristina Selva, de 32 años, el temblor le cogió en casa con sus dos hijas. "Se movía mucho el edificio, me metí con mis niñas debajo de una mesa y esperé a que pasara". El miedo a que todo se repitiera hizo que miles de lorquinos dejaran sus casas con mantas y sillas dispuestos a pasar la noche al raso. **PÁGINAS 14 a 19**

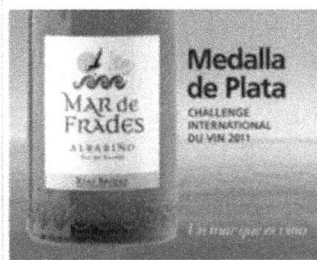

Medalla de Plata
CHALLENGE INTERNATIONAL DU VIN 2011

La presión del PP sobre el 'caso Bildu' lleva la preocupación electoral al PSOE

Rajoy sigue en silencio pero Basagoiti carga contra el Gobierno

C. E. CUÉ, Madrid

El PP golpea duro al PSOE a cuenta del expreso etarra que pidió el voto para Bildu. Entre algunos dirigentes del PSOE hay preocupación porque este es un asunto muy sensible en zonas donde se juegan las elecciones, como Castilla-La Mancha y Extremadura, regiones en las que los socialistas tienen un electorado más conservador y fronterizo que el PP trata de conquistar con estrategias como esta. **PÁGINA 16** **EDITORIAL EN LA PÁGINA 34**

El partido xenófobo impone los controles fronterizos en Dinamarca

R. M. DE RITUERTO, Bruselas

El Gobierno danés anunció ayer el próximo restablecimiento permanente de controles fronterizos con Alemania y Suecia. La decisión, fruto de un acuerdo con el xenófobo Partido Popular Danés, ignora Schengen y se produce cuando la UE se dispone a revisar el tratado. **PÁGINAS 5, 38 y 39**

El clan Asad desplaza al líder sirio y agudiza la represión

ENRIC GONZÁLEZ, Jerusalén

El clan familiar Asad ha desplazado al presidente sirio, Bachar el Asad, y se ha puesto al frente de la represión contra los manifestantes que tienen en jaque al régimen. El Ejército bombardeó ayer la ciudad de Homs, uno de los bastiones de la protesta. La ofensiva en esa ciudad y en Deraa causó una veintena de muertos. **PÁGINAS 2 y 3**

El País

1,20€

elPeriódico de Catalunya

ju. 12

Fem els 5.000 en català

Terremoto mortal en Murcia

▸ Un temblor de 5,2 grados causa al menos 8 muertos, la mayor tragedia sísmica en España desde 1956

Páginas 28 y 29

UN AÑO DESPUÉS

¿Valió la pena?

El tijeretazo de Zapatero ha alejado la amenaza de la intervención europea, gracias al control del déficit, pero ha frenado el crecimiento

Artículos de José Antonio Bueno, José Carlos Díez y Josep Olivé ▸PÁGINAS 2 A 5, 7 Y EDITORIAL

El Periódico

LA GACETA

DE LOS NEGOCIOS. Número 6.807 Jueves, 12 de mayo de 2011. Precio: 1,20€. www.gaceta.es • EL DIARIO DE MAÑANA •

AL MENOS OCHO MUERTOS EN DOS TERREMOTOS Y LA CAMPAÑA ELECTORAL SE SUSPENDE

Durante cinco horas Murcia vivió el pánico de Japón

● Zapatero moviliza al Ejército ● Los temblores se notaron en varias provincias y llegaron al sur de Madrid ● Miles de personas pasaron la noche en el recinto ferial ● La ciudad sufrió otro seísmo hace seis años

ENVIADA ESPECIAL
Belén Rueda
Murcia

El mismo día que se cumplían dos meses del seísmo y el tsunami en Japón, la tierra tembló en Murcia y sus provincias colindantes causando al menos ocho muertos, entre ellos una embarazada y dos niños, y decenas de heridos. Los terremotos, sucesivos, alcanzaron una intensidad de 5,1 grados en la escala de Richter y se llegaron a sentir incluso en el sur de Madrid. El epicentro se localizó en Lorca, donde los vecinos se negaron a regresar a sus casas ante el temor a nuevas réplicas. Al menos 10.000 personas fueron realojadas en un recinto ferial. **Pág. 42 a 45**

En Lorca se produjeron numerosas escenas de pánico tras el terremoto que provocó cuantiosos destrozos.

España, un año después del tijeretazo

Zapatero niega sus recortes **Págs. 24 y 25**

PARADOS (Miles)
1 Trim./11 .... 4.910
II Trim./11 ... 44.640

TASA DE PARO (%)
I Trim./11 21,29
II Trim./10 ... 20,09

IPC (%)
Marzo-2011 3,6
Mayo-2010 1,8

PIB (%)
IV Trim./10 0,2
I Trim./11 0,3

NUEVO E INMORAL DESPILFARRO EN LA ANDALUCÍA DEL PARO Y LA CORRUPCIÓN

Tres consejerías de Griñán se dedican a enseñar 'porno' a los adolescentes

● El programa se llama Forma Joven y va destinado a 200.000 niños de entre 12 y 16 años ● Se imparte ya en 750 colegios públicos y concertados de la comunidad **Pág. 43**

El Barcelona certifica su tercera Liga consecutiva **Pág. 60**

DESAFÍA EL PASO DEL TIEMPO DE MANERA INTELIGENTE

Revidox

STILVID® 84%

El mayor avance científico antiage

acta

www.actafarma.com

En el día de hoy
Una tragedia que nos cambia la faz 12

Los periódicos ayer hemos cambiado de faz. Nuestra primera plana contenía una información sobre uno de los grandes escándalos que afectan a la Junta de Andalucía. A lo largo de la tarde de ayer constatamos que el día no estaba ni para exclusivas diferentes ni para alguna broma electoral como la que tenemos en curso. Murcia ha sufrido una tragedia de la que, cuando escribo, no se alcanza aún a saber la magnitud. El mundo entero y España, por tanto, lleva dos meses exactos conmocionado por la tragedia de Japón. Afortunadamente, la que ayer padeció Lorca no ha alcanzado esos niveles de drama humano. Cuando ocurren estos tremendos episodios naturales caemos en la cuenta de que las cosas que nos ocupan a diario se quedan extraordinariamente pequeñas. Hablar ahora de campañas electorales, por ejemplo, supondría una enorme frivolidad. **Carlos Dávila. Director**

DOMINGO 15 DE MAYO

DVD ESPAÑA EN LA MEMORIA "LAS PLAZAS DE TOROS ANTIGUAS Y LOS MALETILLAS"

POR SÓLO 2€ MÁS

PROMOCIÓN AMPLIADA 5ª ENTREGA MAÑANA PULSERA DE LA VIRGEN DEL CARMEN

por 5,95€ más

La Gaceta

TRIBUNALES @ 22

El juez Juan del Olmo reabre el caso del yacimiento de San Esteban y ordena la imputación del director general

FÚTBOL
Tercer título de Liga del Barça de la 'era Guardiola' @ 50 a 55

HOY CON La**Opinión**
GUÍA DE PARQUES NATURALES
MANUAL DE PEGO-OLIVA

La**Opinión** DE MURCIA

www.laopiniondemurcia.es

JUEVES, 12 DE MAYO, 2011 | AÑO XXIII | NÚMERO 8.260 | DIRECTORA PALOMA REVERTE DE LUIS   PRECIO 1.10 €

El cuerpo de una de las víctimas permanece en el suelo entre cascotes. JUAN CABALLERO

Horror en Lorca

▶ Dos terremotos causan al menos ocho muertos, entre ellos un menor y una embarazada ▶ Los 270 enfermos del Rafael Méndez fueron desalojados

■ Al menos ocho personas murieron ayer en Lorca sepultadas bajo cascotes y piedras que se desprendieron de decenas de edificios tras temblar la tierra. El primer terremoto, de 4,4 grados, sacudió la Región poco después de las cinco de la tarde, pero la réplica que le siguió dos horas después fue la causante de los mayores daños. @ 2 a 16, 36 y última

Unas 20.000 personas pasan la noche en la calle ante el temor a que se produzcan nuevas réplicas

Rubalcaba y Rajoy viajan hoy a la Región después de que se haya suspendido la campaña electoral

4,4

5,2

Los bomberos rescatan a una vecina de entre los cascotes. XXXXXXXXXX

La Opinión

LA RAZÓN

DIARIO INDEPENDIENTE DE INFORMACIÓN GENERAL · JUEVES 12 de mayo de 2011 · Año XIV · 4.535 · PRECIO 1,20 euros · EDICIÓN MADRID

Hoy, gratis, nuevo cristal para las gafas

Terremoto mortal en Lorca

Dos seísmos consecutivos causan al menos ocho muertos y decenas de heridos en Murcia PRIMERA PLANA, P. 16

El terremoto provocó escenas de pánico y dolor en la ciudad murciana. A la izquierda, una de las víctimas

Bildu obtuvo créditos de más de dos millones antes de su legalización

- Solicitaron el préstamo tras su constitución, el 3 de abril, dando por hecho que lograrían estar en las elecciones

- Una parte del dinero se ha gastado en dar a conocer la coalición y el resto lo destinarán a la campaña P. 24

MATTHEW PERRY

MR. SUNSHINE

ESTRENO EN **EXCLUSIVA**

ESTA NOCHE **20:40h**

SONY ENTERTAINMENT TELEVISION

canalsonytv.es

Digital+(dial 17), Euskaltel (dial 21), Movistar Imagenio (dial 32), Orange TV (dial 13), R (dial 30)

La Razón

LA VANGUARDIA

Tercera Liga del Barça de los prodigios

La tierra sacude Lorca

Al menos 8 muertos en dos seísmos de magnitud 4,4 y 5,2

Diez mil personas pasan la noche fuera de su casa

El terremoto daña muchos edificios de la ciudad murciana

'Aquí en el campo no puede caerme nada encima'

Ayuda catalana en el rescate

La Vanguardia

LA VERDAD

ALICANTE
ELCHE
Jueves, 12-05-11
Nº 34.380 • 1,20€

FUNDADO EN 1903. PERIÓDICO DE LA PROVINCIA DE ALICANTE www.laverdaddealicante.es

JAPÓN, DOS MESES TRAS EL DESASTRE — Bajo la amenaza nuclear, los japoneses recuperan la normalidad **P63**

EL ALTET ALCANZARÁ ESTE AÑO LOS 10 MILLONES DE VIAJEROS **P20**

Un seísmo lleva la tragedia a Lorca

Al menos ocho muertos y dos heridos muy graves en la ciudad murciana por el terremoto con más víctimas desde 1956 **P36-38**

La hija de uno de los fallecidos es consolada inmediatamente después del terremoto, al fondo, el cuerpo de otra de las víctimas. :: ISRAEL SÁNCHEZ / EFE

Bienestar Social bloquea 9.000 ayudas a dependientes pese a estar aprobadas

Familiares de afectados acusan al Consell de paralizar la ley y el Síndic recibe un alud de quejas

La consellera del área exige al Gobierno que financie la norma «como corresponde» **6.A. P13**

EL HÉRCULES YA ES EQUIPO DE SEGUNDA **P52-55**

EL BARCA CANTA EL ALIRÓN TRAS EMPATAR **P56**

EMOTIVA DESPEDIDA A SEVERIANO BALLESTEROS **P60**

'CASO GÜRTEL'

El TSJ permite que declaren el escolta y el chófer de Camps
C.M.A. P20

22-M

Bono defiende que es posible una solución al Tajo-Segura
J.C.R. P5

ALICANTE

Plan de Adif para evitar daños en edificios por el soterramiento
T.R. P14

ELCHE

La empresa reanuda las obras en el museo de La Alcudia
J.C.R. P60

Juntos Podemos *crear empleo*

Vota Mercedes Alonso

Centrados en ti

PP

HOY,
VAJILLA ROBERTO TORRETTA PLATO POSTRE, SÓLO 1,60 €

La Verdad

EL DIARIO DE LA COMUNITAT VALENCIANA DESDE 1866. www.lasprovincias.es

VALENCIA

LAS PROVINCIAS

TERROR EN LORCA

Dos terremotos causan al menos ocho muertos y siembran el pánico en la población

Unas 10.000 personas pasan la noche al aire libre en el recinto ferial

Zapatero y Rajoy acuerdan suspender los actos de campaña de hoy

«Yo estoy con Camps a muerte, nunca haría nada contra el PP»

22M Alfonso Rus dice que pudo equivocarse en la exposición del MuVIM

Mantiene que no tiene más aspiraciones en el partido

El presidente del PP valenciano reafirma su apoyo a Camps, dice que nunca haría nada en contra de su partido y niega mayores aspiraciones. Reconoce que él pudo equivocarse con la polémica del MuVIM. R. PALOMAR P20-21

LA ABSTENCIÓN AMENAZA UNA CAMPAÑA ABURRIDA

DÍA 7 CRÓNICA DE H. ESTEBAN

VALENCIA
El turismo pide comercios abiertos los festivos en áreas estratégicas

VALENCIA
Dos familias admiten a la fiscal que compraron a sus hijas

CULTURAS
El Palau de les Arts recibe un aluvión de interesados en celebrar su boda

SAN NICOLÁS DE BARI

BARÇA CAMPEÓN Y LEVANTE DE PRIMERA

Guardiola logra su noveno título en tres años el día en que el Hércules baja a Segunda

EL VALENCIA EMPATA Y NO ATA EL TERCER PUESTO

CONSIGUE MAÑANA EL PRIMER CUPÓN
LAS PROVINCIAS
SUPERMERCADO

Las Provincias

Levante

Diana Gonçalves

Hurricane Katrina: Contesting Singularity in *Treme*

Hurricane Katrina[1] hit the US coast in 2005, striking southeast Louisiana on August 29 as a category 3 hurricane and causing massive destruction. The place most devastated was New Orleans, not only as a direct result of the strong winds and heavy rain but particularly due to floods caused by levee breaches. Given the fact that New Orleans is mostly below sea level,[2] the failure of the levee system that protected the city from surrounding canals and lakes led to a catastrophic outcome: about 80% of the city became flooded, property damage was huge, the number of dead and missing people as well as evacuees skyrocketed. According to the National Hurricane Center, Katrina was the third deadliest hurricane of all time in the United States, taking the lives of around 1,200 people, behind only the Galveston hurricane in Texas and the Okeechobee hurricane in Florida. It was also the costliest natural catastrophe to ever affect the United States, costing around $108 billion dollars and far more than hurricanes such as Ike or Andrew.[3] Moreover, Katrina also brought with it an enormous psychological cost to Americans, especially to New Orleanians, who felt ashamed and humiliated by their incapacity to handle the catastrophe and prevent the levee problem.[4]

---

1 The tradition of naming hurricanes in the United States took its first steps during World War II to simplify the tracking process of multiple storms emerging and moving about the Atlantic and Pacific Oceans. However, it was not until 1953 that the National Weather Service officially began using names for storms. Through to 1979, however, it used only female names. The naming of the storms is now controlled by an international committee of the World Meteorological Organization, which establishes a list of names for Atlantic hurricanes that is used in rotation and recycled every six years. If a storm is too deadly or too costly, such as Katrina (and others mentioned in this paper, e.g. Betsy, Andrew or Ike), the name is removed from the list and another is chosen to replace it.

2 In line with this, New Orleans is usually likened to a giant bowl, delimited by natural barriers and manmade defenses from the waters of Lake Pontchartrain to the north and the Mississippi River to the south and west.

3 See the NOAA [National Oceanic and Atmospheric Administration] Technical Memorandum NWS NHC-6 (2011) by Eric S. Blake, Christopher W. Landsea and Ethan J. Gibney.

4 A problem the Dutch share but, nevertheless, mastered a way to coexist with it a long time ago. Like New Orleans, the Netherlands is largely below sea level and has throughout history been continuously faced with the threat of severe floods. As a consequence, the Dutch were forced to design and build an efficient levee system to protect the most vulnerable areas. The Dutch system has frequently been pointed out as an example for New Orleans to follow. See,

1 *Treme*: The Afterlife of Catastrophe

The question of the psychological costs inherent to ineffective prevention and belated and overdue action provides one of the key topics in *Treme*, the television series by David Simon and Eric Overmyer which premiered on 2 April 2010. The series is set in New Orleans and is broadly about the working-class neighborhood of Treme, Faubourg Treme or Treme/Lafitte as it is often called, recalling its French roots. The neighborhood of Treme is located in the Mid-City District area, is one of the oldest neighborhoods in New Orleans and characterized for being a place where American, African, and Creole cultures come together. Starting three months after the hurricane, *Treme* is thus about the post-Katrina New Orleans. It showcases a group of locals – from a college professor to an attorney, a chef, a bar owner, musicians, a Mardi Gras Indian chief, among others – trying to pick up their lives after the catastrophe while New Orleans is still trying to put itself back together amidst corruption, incompetence, and neglect. The series presents a city still mourning for its dead, grieving for the uncountable still missing and waiting for the return of those who escaped and have not yet returned. The first season, consisting of 10 episodes, is therefore about how to survive catastrophe and its aftermath.

The character I focus on is Creighton Bernette, a college professor at Tulane University and a novelist, played by John Goodman, who embodies the anger and anguish of New Orleanians. Creighton acts as the voice of the city, repre-

---

for example, the article by US environment correspondent Suzanne Goldenberg for *The Guardian* where she describes how America has been forced "to abandon the outmoded 'patch and pray' system of levees – whose failure magnified the devastation of Hurricane Katrina – and borrow from the Dutch model of dykes and water management" (2009). See also "Physical Constraints on Reconstructing New Orleans", where Robert Giegengack and Kenneth R. Foster analyze the relationship between New Orleans and the Netherlands, focusing however on the differences between the flood control systems adopted by the two: "While the magnitudes of investment in the Delta Works project in the Netherlands and in a proposed project to protect New Orleans against a Category 5 storm might be comparable, there are very large differences. One is in the level of protection. The Dutch system was designed to offer a 10,000 year protection level, while a Category 5 storm in the vicinity of New Orleans would be roughly a 1 in 200 year event. In important respects, the Dutch have a simpler task: Amsterdam does not lie on a coast subject to the kinds of storm surges routinely associated with tropical storms, and Amsterdam faces a potential enemy on only one front: it does not have one of the world's largest rivers carrying 550,000 Mcm/yr through the heart of the city on an elevated trackway. There is also a huge difference in the economic value of the assets protected – in the case of the Netherlands, the whole country is at risk, while in the case of New Orleans, a similar investment would be required to protect a single city" (2006, 29–30).

senting on the one hand, the inhabitants' frustrations through passionate speeches and rants full of obscenities and, on the other, the search for solutions through simple common sense. Creighton becomes aggravated every time he is confronted by the argument that the Katrina flooding was a natural event. He vigorously defends that the levee breaches and the subsequent flooding result- ed from human error, specifically by the federal government (the Army Corps of Engineers) who projected and built the levee defenses. One of the most em- blematic scenes is when Creighton argues with a British reporter affirming that what hit the Mississippi Gulf coast was in fact a natural disaster; with the flood- ing of New Orleans nevertheless being a manmade catastrophe:

Reporter: Are you saying this was a natural disaster, pure and simple?
Creighton: A natural disaster?
R: A hurricane is.
C: What hit the Mississippi Gulf Coast was a natural disaster, a hurricane, pure and simple. The flooding of New Orleans was a manmade catastrophe, a fed- eral f----up of epic proportions, and decades in the making. [the city, repre- senting lown]. Not in '65 and not three months ago. The flood protection sys- tem built by the Army Corps of Engineers, aka the federal government, failed. And we've been saying for the last 40 years, since Betsy, that it was gonna fail again unless something was done. And guess what? It was not. [...] The levees weren't blown. The floodgates failed, the canal walls failed, the pumps failed. All of which were supposedly built to withstand a much great- er storm (Ep. 1).

The reasoning behind this type of discourse appears to uphold even today as demonstrated by President Barack Obama's Katrina anniversary speech at Xavi- er University in 2010. The speech is in accordance with the statements above and also describing Katrina as a manmade catastrophe. President Obama's words could thus be perfectly attributed to Creighton, although the latter would have not proffered them in such a gentle manner. President Obama declared:

It's been five years since Katrina ravaged the Gulf Coast. There's no need to dwell on what you experienced and what the world witnessed. We all remember it keenly: water pouring through broken levees; mothers holding their children above the waterline; people stranded on rooftops begging for help; bodies lying in the streets of a great American city. It was a natural disaster but also a manmade catastrophe – a shameful breakdown in gov- ernment that left countless men and women and children abandoned and alone (*The Times Picayune* 2010).

In both cases, a clear distinction is drawn between the damage directly caused by the hurricane and the devastation generated by human negligence which ushered a devastating outcome, both physically and socially.

Katrina is not, however, the only catastrophe addressed by Creighton. Throughout the first season, Creighton, who dies in episode 9 when he jumps from a ferryboat into the Mississippi river, thus emulating the death of his beloved city, is constantly finding interlinking elements between Katrina and other catastrophes from the past, such as the Great Mississippi River Flood of 1927, which, until Katrina, was the greatest natural catastrophe in U.S. history. One of the reasons for the frequent summoning of previous catastrophes is the fact that for several years Creighton has been working on a book about the Great Mississippi Flood. The post-Katrina period strikes as the perfect time to take his book out of the drawer since, as he admits himself, it "couldn't be more topical" (Ep. 2). This is exactly the logic deployed by his publisher to persuade him to write about Katrina. Six months after the hurricane, his publisher has regained interest in his book and in Creighton's role as spokesman for New Orleans. Trying to seize the opportunity, Creighton's agent suggests the inclusion of references to Katrina and rushes him to finish the book. Creighton, nevertheless, refuses the idea of changing his subject and promises a manuscript within six weeks:

Creighton: Carla, I thought Random House wrote me off a long time ago.
Agent: Maybe they did, to a certain extent, but Katrina has put New Orleans back on the map in a certain way.
C: It nearly wiped it off the map, actually. The near-death of an American city and now my publisher's hot for a book they couldn't care less about six months ago.
A: Actually, they're not only interested in bringing your novel out on next fall's list, but they want you to do something contemporary before then.
[...]
A: [I]f you can manage to make the novel contemporary, maybe set part of it in post-Katrina ...
C: The novel is about the 1927 flood. It speaks to the present situation only as metaphor.
A: Well, perhaps if you could bookend some older material ...
C: Jesus Christ, Carla. I thought you said they wanted the book.
A: They do, but in the voice that you have become. You're a spokesman now for this city.
C: I'm not trying to be the spokesman of the city. New Orleans speaks for itself. And the novel is what it is. If they still want it, fine. (Ep. 6).

As we come to know Creighton Bernette, we realize however that he is struggling with writers' block. After Katrina, he finds it more difficult to get back to the writing of his book, considering his fiction too inconsequential in comparison to the reality he is living (Ep. 4). Tired, angry and uninspired to write his fiction, Creighton turns to the Internet as a platform to showcase his take on

Katrina, New Orleans, the government, and the abandonment of the city by the rest of the country. Creighton Bernette's character was in fact loosely based on the late blogger Ashley Morris, who died in 2008 and was famous for his blog about the state of New Orleans and the disregard of governmental institutions. On 25 November 2005 (3 months after the hurricane), Ashley Morris posted on his blog what came to be known as his most acclaimed post and later used as inspiration for the series. In that post, Morris wrote about the need for the rebuilding of New Orleans, giving it the same chance of surviving the catastrophe as had been given before to other cities. Morris wrote: "What about you f---s that don't want to rebuild NOLA because we're below sea level. Well, f---heads, then we shouldn't have rebuilt that cesspool Chicago after the fire, that Sodom San Francisco after the earthquakes, Miami after endless hurricanes, or New York because it's a magnet for terrorists" (2005).

Creighton Bernette's rants on YouTube, although they do not reproduce Morris's words in their entirety, maintain the same spirit of criticism and resilience. They aggressively criticize those who question the need for helping all New Orleanians affected by the hurricane and the subsequent floods and eventually rebuilding the city, downplaying to an extent the cultural and historical significance of New Orleans. Therefore, it can be said that these rants reflect, above all, the disappointment and fatigue felt by New Orleanians on the one hand, and all the comparisons that emerged during the post-Katrina period on the other:

> You say, "Why rebuild it?" I say, "F--- you." You rebuilt Chicago after the fire. You rebuilt San Francisco after the earthquake. Let me tell you something. Anything that's any f---ing good in Chicago came from someplace else. And San Francisco is an overpriced cesspool with hills. [...] To New York, f--- you too. You get attacked by a few fundamentalist f---ing assholes and the federal money comes raining down like rose petals. Our whole f---ing coast was destroyed and we're still waiting for somebody to give a good goddamn. (Ep. 4).

2 Looking Back in History: Contesting Singularity

Throughout the whole season, Creighton's rants are a glimpse of a trend usually present in responses to catastrophes: the reference to previous catastrophic moments so as to make sense of what is happening. In order to prove the validity of his argument that Katrina was an unnatural natural disaster, Creighton uses examples both from natural and manmade catastrophes. Furthermore, the reference to a number of catastrophes that are handled and represented by means of a repeated structure highlights the importance of preserving catastrophe in cultural memory. Catastrophes are the perfect opportunity to learn from the past and create something new, which brings to mind the Benjaminian conception of progress as a product of catastrophe.[5]

Creighton recalls the Great Chicago Fire of 1871, the 1906 San Francisco earthquake, the Mississippi Flood of 1927, Hurricane Betsy of 1965, or a more recent catastrophe, 9/11. These specific examples are representative of the fascination with urban catastrophes, which render the vulnerability of cities to catastrophic events visible. Cities encompass two factors that fuel catastrophe, amplifying its impact: their population density and their extremely high concentration of man-built structures. A catastrophic event in a city necessarily affects at least one of these facets and usually on a large scale. Occurring ever since the biblical Sodom and Gomorrah, or even Pompeii and the Lisbon earthquake, Katrina is but the most recent example of a familiar theme that is frequently replayed in history. Additionally, it is also interesting to realize that the examples presented by Creighton cover a wide spectrum of catastrophe types – fires, earthquakes, floods, hurricanes, and terrorist attacks – that comprise the four elements: fire, earth, water and air. Moreover, they also represent catastrophic moments where natural catastrophe intertwines with human action. Katrina blended natural and technological disaster and has been perceived as a "natech disaster".[6] The examples given by Creighton have the particularity of, at a certain point and to some extent, combining natural and technological features: they are events where the actions of nature and men are at times blurred and natural disasters escalate due to human error or unpreparedness; or even manmade disasters whose effects are intensified by the force of nature.

I believe the main goal in Creighton providing such a wide range of examples lies originally in the intention of affirming the singularity of Katrina. In comparing it to various catastrophes and, in the end, finding them always wanting, the focus remains on Katrina's uniqueness and incomparable quality. However, what should be questioned here is not Katrina's singularity *per se* but the singularity of its reception. Taking 9/11 as a model, after catastrophic and trau-

---

5 See Walter Benjamin, *The Arcades Project*: "The concept of progress must be grounded in the idea of catastrophe" (N9a,1).

6 Both geographical location and an easygoing attitude of governmental institutions and inhabitants, who overlooked all the risks intrinsic to News Orleans (a city mainly below sea in a hurricane-prone area and equipped only with an obsolete and feeble prevention system) were determining factors to the destructive power of Katrina. See, for example, Picou, "Katrina as a Natech Disaster Toxic Contamination and Long-Term Risks for Residents of New Orleans" (2009), or Picou and Marshall, "Katrina as Paradigm-Shift: Reflections on Disaster Research in the Twenty-First Century" (2007) to better assess the impact of Katrina as a natech disaster.

matic events, people usually turn to the past in order to better read the present, find a suitable narrative and, thence, construct the story of the newest catastrophe in a different light. With 9/11, references to the Lisbon Earthquake, Pearl Harbor, Hiroshima, or the explosion of the *Challenger* were almost immediate. From this point of view, to those more historically aware, comparisons are a natural process of turning to the past in order to better respond to the new and the strange through a recognizable frame. They are, in the words of Elaine Tyler May, a way "to meet the unfamiliar with the familiar" (2003, 36).

In the case of Katrina, the references to hurricanes and floods serve both as reassurance (that something similar had already happened and could be overcome) and as a decisive factor against the naturalness or inevitability of the event (proving that, if one had learned from the past, the present catastrophe could have been avoided or at least attenuated). Likewise, references to other kinds of catastrophes, from fires to earthquakes or even terrorist attacks, are a means of defying the traditional distinction between natural and technological catastrophes by reviving circumstances that coupled characteristics of both. However, this also proves a way of bearing in mind the notion that catastrophes, no matter which, are part of history and collective memory and that the way people respond to a new event is always influenced and conditioned by how it has been done in the past.

It is my contention that the examples given by Creighton can be separated into two categories, one related to the catastrophic moment itself and the other to its aftermath. I divide them into: 1) the search for a common bond; 2) the need for reconstruction. As far as the first is concerned, this relates to the desire to look to history for events that share similar characteristics and, thus, that may help the (con)textualization of the newer catastrophe. What this means is that by looking into past catastrophic events that had an analogous origin or impact, people try to place and inscribe the newest catastrophe in history and cultural memory as well as interpret what they have witnessed so as to build the story of the event, i.e. that which will be remembered by future generations. In relation to this, in her essay about the visualization of catastrophe in Walter Benjamin and Spike Lee, more specifically regarding the Great Mississippi Flood and Katrina, Isabel Gil (2011) stresses the fact that what future generations will know about a catastrophe is not exactly what happened but how it was visualized, how it was mediated.

Two moments are revisited here: the Mississippi River Flood of 1927 and Hurricane Betsy of 1965. Very briefly, the Great Mississippi River Flood resulted from heavy rains for a long period of time and the subsequent rise in the tide. Due to extremely high water levels – levels that remain a record to this day – the Mississippi broke out of its levee system, inundating an immense area. The specific case of New Orleans was somewhat different since the levees did not naturally break but were instead destroyed with dynamite to divert water from the city into low-lying land, namely the St. Bernard and Plaquemines Parishes.[7] As far as Hurricane Betsy is concerned, it made landfall in Louisiana as a category 3 hurricane. Having formed in Florida, it entered the Gulf of Mexico, reaching Lake Pontchartrain and the Mississippi River, where it generated huge waves. Levee breaches led to the flooding of New Orleans, killing dozens of people and turning Betsy into the first hurricane to top $1 billion dollars in damages, thus earning the nickname "Billion-Dollar Betsy". Due to the similarities to Katrina and its effects, especially in terms of the failure of the levees and the subsequent floods, these two examples were repeatedly recalled in the aftermath of Katrina. Notwithstanding their intrinsic differences, given the fact that these are always situated events – influenced by the social, historical, political and economic context in which they occur –, both the Mississippi Flood and Betsy functioned as models for the representation of Katrina (recapturing a common imagery, like that of flooded streets and houses); but also for the political response (passing legislation and adopting instrumental measures to prevent and control floods). Accordingly, the Mississippi Flood led to the Flood Control Act of 1928, which stated that the Army Corps of Engineers was in charge of improving the levee system to prevent future floods of the magnitude of that of 1927; the same type of response was adopted after Betsy with the Flood Control Act of 1965. Once more, the destruction wrought by Katrina also resulted in the rehabilitation, replacement or construction of more levees. In addition, after Katrina, in 2008, the National Response Framework was set up, presenting guidelines for a national response to catastrophes or emergencies. This Framework replaced the National Response Plan of 2004, which had come into effect after 9/11.

Catastrophes generally result in the implementation of new political and social measures. They force policies to be reconsidered and new procedures to be introduced so as to prevent the same thing from happening again or at least to reduce its impact. This is related to the idea of reconstruction, which also entails the notions of restructure, reformulation, and recovery. Part of the reconstruction process occurs physically in the space affected by the catastrophe. This serves as the perfect bridge to the second category: the need for reconstruc-

---

7 *Rising Tide: The Great Mississippi Flood of 1927 and how it changed America* by John M. Barry is a pivotal narrative of the events of the Great Flood, thoroughly exploring its social impact. This book has witnessed a renewed interest after Katrina due to the similarities between the catastrophes of 1927 and 2005. Justifying the importance of this book to the study of the Great Flood, in one scene from *Treme*, Creighton appears reading it (Ep. 3).

tion.[8] This category departs from the ideas of destruction and (re)construction. Catastrophes are a symbol of destruction, devastation, and decay. They are closely connected with, on the one hand, the notions of fall and ruins; on the other hand, however, they are also linked with the understanding that destruction implies the erasure of something and the need for starting anew. In his first YouTube rant, Creighton Bernette called attention to the tradition of facing ruin through reconstruction. Enumerating various moments in US history where the devastation caused by catastrophe led to the rebuilding of the space, Creighton underlined the importance of rebuilding New Orleans as well, contradicting the argument that frequently emerges after such a devastating and perhaps preventable event that some disaster-prone locations should simply be abandoned.[9]

Concentrating now solely on the examples presented by Creighton, the Great Chicago Fire, the San Francisco Earthquake and 9/11 are briefly explored below and following the same pattern as the catastrophes previously analyzed. The Great Chicago Fire was an enormous fire that lasted two days and destroyed the whole city of Chicago. Due to precarious construction methods, which consisted of the use of wooden structures, the fire spread rapidly. Chicago represents the materialization of the idea of catastrophe as progress. Given the extent of the damage, the city had to be rebuilt from scratch, giving rise to new architectural styles and the use of new techniques and materials, such as the use of steel for the frames of the buildings (allowing for thinner walls, more windows and more height). In fact, Chicago became home to the world's first skyscraper, the Home Insurance Building, which was constructed after the fire in 1884.

---

8 In his essay "Restoring Urban Viability" (2006), Lawrence J. Vale explores how cities recover from disaster. The author examines for example how government promises a better than ever reconstruction so as to compensate for destruction; how the decision of which part of the population is brought back contributes to the dominant narrative; the efforts made to reconcile population and governmental institutions; or the restoration of landmarks to work as a provider of a sense of renewal and wellbeing on the one hand, and as a distraction from the rest on the other.

9 Facing ruins also means facing the past since our perspective is inevitably framed by a long tradition of catastrophes and ruin gazing. In the introductory chapter of *Ruins of Modernity*, Julia Hell and Andreas Schönle argue that images of destruction always resonate with other images of destruction, thus evoking "a transhistorical iconography of decay and catastrophe, a vast visual archive of ruination" (2010, 1). Moreover, according to the authors, America has always approached ruins very differently from Europe. Americans usually did not pay much attention to them, preferring to value their natural heritage. 9/11 however constituted an important turning point in the way of perceiving ruins. This discourse is recaptured by Creighton in a conversation with his daughter Sofia when he takes her for a walk to the southern shore of Lake Pontchartrain and, looking at all the destruction and reminiscing about what that area used to be like, bitterly complains: "It's like seeing an ancient ruin" (Ep. 8).

In 1906, the San Francisco Earthquake struck San Francisco and other areas in Northern California along the San Andreas Fault. Throughout the nineteenth century, San Francisco had been growing fast, becoming the largest city in California in the process. Nevertheless, although most constructions ultimately incorporated some earthquake protection measures along with resistance to preventing the spread of fires, some developers built in unstable regions, filling land that was susceptible to liquefying during a large earthquake. When the earthquake happened, structures on such filled land were the most damaged while the new steel-frame buildings were those most unaffected by the quake. However, the subsequent fire that spread throughout the whole city eventually destroyed what the earthquake had spared. The need for quick reconstruction was emphasized almost immediately. On the one hand, San Francisco wanted to keep its place as the economic leader of the western U.S., on the other hand, it wanted to seize the opportunity to restructure the city not only in terms of creating distinguishable districts but also in terms of taking care of pending social issues, such as trying to move the Chinese population to the outskirts of the city.

Finally, regarding 9/11, as a result of the attacks and subsequent fall of the towers, the World Trade Center site, or Ground Zero as it became known, was turned into ruins, into dust and debris. Although it may seem contradictory, applying the expression Ground Zero to designate the World Trade Center site already denotes a desire for reconstruction. Originally, Ground Zero is linked to an explosion scenario thus evoking images of a nuclear detonation, such as that of Hiroshima. Nevertheless, the idea of complete devastation also allows for the beginning of something new. In this sense, Ground Zero functions both as the point of the most acute destruction but also as a point of departure (point zero). Contrary to what happened in New Orleans after Katrina, as is patent in Creighton's words, the insistence to rebuild New York contrasts enormously with the negligence experienced in the reconstruction of New Orleans. Since 2001, work of enormous proportions and social impact has been done to recover the area affected by the attacks. The works have involved, for example, the construction of office buildings, a Performing Arts Center and a Memorial that symbolize the attempt to both reclaim what the former World Trade Center stood for and, conversely, materialize the need to never forget what happened. Taking all this into account, the reconstruction of devastated spaces can operate as a reminder of catastrophe. Even though reconstruction may promise the erasure of the old, catastrophe remains nevertheless present in the space. One can never ignore the always-latent relation between what used to be and what is. Reconstruction implies the idea of a new beginning and especially that there is a before and an after. From this perspective, the process of physical (re)construction involves a mental (re)construction as well, leading people away from suffering and trauma, working as tools of mourning, healing, and closure.

It can thus be said that Katrina resonates with images, narratives and measures from past catastrophic events. What took place in New Orleans in 2005 turns out to be a mixture of an unexpected event with an event heir of past experiences.[10] Indeed, in his attempt to heighten Katrina's singular nature, Creighton Bernette does exactly the contrary. By evoking past catastrophes, he inscribes Katrina in a group of events that, although different and unique, still share important features and response mechanisms. Against this backdrop, it can be said that representation of catastrophe breaks with singularity. An event may be perceived as singular, i.e. incomparable or without equivalent in its socio-historical contingency, but that does not exclude the possibility of comparability on the reception end. According to Derek Attridge, singularity is an event that occurs in reception. The singularity of a cultural object is generated by a configuration of properties that transcend pre-programmed norms. This singularity however is always open to reinterpretation and recontextualization (2005). This definition of the singular as a constantly changing event produced in reception moves in the same direction as the argument that I am trying to build regarding the representation of catastrophe as a non-singular event. A catastrophe may present singular or extraordinary features, yet its representation brings forward the realization that there is a pre-existing iconology or narrative, a previous rhetoric for dealing with these distinctive events.

From this point of view, everything we see, say and write about Katrina (or any other catastrophe as a matter of fact) is but some kind of rewriting, of retelling, of reconstructing a different reality, sharing the fact of remediation as Jay Bolter and Richard Grusin would say (2000). However, this process works both

---

10 As we can already sense from Creighton's words, the idea of unexpectedness in relation to Katrina has been widely criticized by many scholars. In fact, already in 2001 (reprinted in 2005), Eric Berger in a *Houston Chronicle* article had pointed out: "New Orleans is sinking. And its main buffer from a hurricane, the protective Mississippi River delta, is quickly eroding away, leaving the historic city perilously close to disaster. So vulnerable, in fact, that earlier this year the Federal Emergency Management Agency ranked the potential damage to New Orleans as among the three likeliest, most catastrophic disasters facing this country. The other two? A massive earthquake in San Francisco, and, almost prophetically, a terrorist attack on New York City. The New Orleans hurricane scenario may be the deadliest of all." Also in 2001 (reprinted in 2008), Mark Fischetti in an article for *Scientific American* made the alarming call to attention: "New Orleans is a disaster waiting to happen." Later in 2004, Joel K. Bourne, Jr. in an article for *National Geographic Magazine* explored the impact of hurricanes on New Orleans. Quoting University of New Orleans geologist Shea Penland concerning a disastrous hurricane in New Orleans, the same belief comes forth: "It's not if it will happen. [...] It's when."

ways. If catastrophic events remediate each other; it is also true that, looking back, we may find certain elements that provide a template to deal with the most recent experience, i.e. elements that premediate the event taking place in the present. Therefore, I do not apply here the concept of premediation in the exact same sense as Richard Grusin (2004; 2010), who envisages it as a kind of prevention, anticipation or planning for the future; I prefer to use it in the sense of looking for processes of mediation in the past that may constitute a kind of support or template for coverage of a more recent catastrophe. Grusin's take on this concept is significantly limited by the post-9/11 climate, thus emphasizing "the production of specific future scenarios and the creation and maintenance of an affective orientation towards the future, a sense of continuity or the feeling of assurance that there will not be another catastrophic surprise" (2010, 48). However, as Katrina has proven, 9/11 did not prevent people from once again experiencing shock, surprise, fear, anxiety or trauma. Hence, my interpretation of the concept is in line with that more comprehensively presented by Astrid Erll, who bridges remediation and cultural memory studies and claims that "[t]he term premediation draws attention to the fact that existent media which circulate in a given society provide schemata for future experience and its representation" (2008, 392).

In relation to this idea of a mediated experience, as a result of his analysis of American culture, Kevin Rozario asserts that "we live in a culture of calamity" (2007, 1). In this book, the author investigates American culture and its compulsion for spectacles of calamity, which he depicts as spectacles that "command our attention because they present an occasion for processing, intellectually and emotionally, the experience of living in a world of systematic ruin and renewal, destruction and reconstruction, where technological and environmental disasters always loom (6).

Recognizing the social and cultural relevance of catastrophes, the media have explored and exploited the intrinsic need to let oneself be involved by the spectacle of catastrophe, transforming it and fitting it into the great umbrella of "media events" as Daniel Dayan and Elihu Katz put it (1994). *Treme* seems to follow this logic of compulsion for spectacles of calamity/catastrophe, representing the afterlife of Katrina and activating the memory of previous catastrophic moments through small references made by Creighton Bernette. In the lack of actual images, these references function almost like snapshots of those past catastrophes in the viewers' minds. Viewers may not have an image they can associate with the current one or may not know much about the catastrophes – i.e. they may not have the memory of those events –, but they have the imagination of that memory (framed by representational tropes). This idea gains the support of Marc Redfield who, in *The Rhetoric of Terror*, argues that a "society of spectacle is necessarily an intensely if narrowly verbal society" (2009, 13) and also how it is not just as an "array of images" that a catastrophe becomes part of cultural life but above all "as a name": e.g. The Great Mississippi Flood, 9/11, or Katrina. The name alone in a single expression encloses everything that composes the event and thus playing a paramount role in the way the catastrophe is represented and perceived.

3 Conclusion

To conclude, the perception of a new event is always contaminated (by the perception of previous events, by the preexisting circulation of rhetorical modes of representation). The past informs the present and vice-versa. Events, and especially disruptive events, are not immediately and entirely grasped: they need to be contemplated and to be put in relation to other events for their meaning to emerge. This is what Creighton Bernette does in *Treme*. Both the recovery of the flood imagery and the circulation and repetition of common political strategies for dealing with catastrophe are expected to help produce and structure the representation and reception of Katrina. Taking all this into consideration, we may state that catastrophes are deeply troped: there are images and narratives that one expects to see emerge whenever a catastrophe strikes. As such, Katrina is anchored in a long tradition of a culturally, historically and socially constructed representation of catastrophe, where preexisting frameworks are recaptured whenever a new catastrophe takes place.

References

Attridge, Derek. *The Singularity of Literature*. London: Routledge, 2005.
Barry, John M. *Rising Tide: The Great Mississippi Flood of 1927 and how it changed America*. New York: Touchstone, 1998.
Benjamin, Walter. *The Arcades Project*. Ed. Roy Tiedemann. Trans. Howard Eiland and Kevin McLaughlin. Cambridge and London: The Belknap Press of Harvard University Press, 2002.
Bolter, Jay David, and Richard Grusin. *Remediation: Understanding New Media*. Cambridge: MIT Press, 2000.
Dayan, Daniel, and Elihu Katz. *Media Events: The Live Broadcasting of History*. Cambridge: Harvard University Press, 1994.
Erll, Astrid. "Literature, Film, and the Mediality of Cultural Memory." In *Cultural Memory Studies: An International and Interdisciplinary Handbook*, eds. Astrid Erll and Ansgar Nünning, 389–398. Berlin: de Gruyter, 2008.

Giegengack, Robert and Kenneth R. Foster. "Physical Constraints on Reconstructing New Orleans." In *Rebuilding Urban Places After Disaster: Lessons from Hurricane Katrina*, eds. Eugenie L. Birch and Susan M. Wachter, 13–33. Philadelphia: University of Pennsylvania Press, 2006.

Gil, Isabel Capeloa. "Os Senhores da Imagem. Walter Benjamin e Spike Lee." In Isabel Capeloa Gil, *Literacia Visual. Estudos sobre a Inquietude das Imagens*, 81–95. Lisbon: Edições 70, 2011.

Grusin, Richard. "Premediation." *Criticism* 46, no. 1 (2004): 17–39.

Grusin, Richard. *Premediation: Affect and Mediality after 9/11*. New York: Palgrave Macmillan, 2010.

Hell, Julia, and Andreas Schönle. "Introduction." In *Ruins of Modernity*, eds. Julia Hell and Andreas Schönle, 1–14. Durham: Duke University Press, 2010.

May, Elaine Tyler. "Echoes of the Cold War: The Aftermath of September 11 at Home." In *September 11 in History: A Watershed Moment?*, ed. Mary L. Dudziak, 35–54. Durham: Duke University Press, 2003.

Picou, J. Steven. "Katrina as a Natech Disaster Toxic Contamination and Long-Term Risks for Residents of New Orleans." *Journal of Applied Social Science* 3, no. 2 (2009): 39–55.

Picou, J. Steven and Brent K. Marshall. "Katrina as Paradigm-Shift: Reflections on Disaster Research in the Twenty-First Century." In *The Sociology of Katrina: Perspectives on a Modern Catastrophe*, eds. D. Brunsma, D. Oversvelt, and J. Steven Picou, 1–20. Lanham, MD: Rowman and Littlefield, 2007.

Redfield, Marc. *The Rhetoric of Terror: Reflections on 9/11 and the War on Terror*. New York: Fordham University Press, 2009.

Rozario, Kevin. The Culture of Calamity: Disaster and the Making of Modern America. Chicago: University of Chicago Press, 2007.

Vale, Lawrence J. "Restoring Urban Viability." In *Rebuilding Urban Places After Disaster: Lessons from Hurricane Katrina*, eds. Eugenie L. Birch and Susan M. Wachter, 149–167. Philadelphia: University of Pennsylvania Press, 2006.

Series:

Simon, David and Eric Overmyer. *Treme*. USA: HBO, 2011.

Internet:

Berger, Eric. "A Deadly Disaster in New Orleans." Trans. Howard Eiland. *Houston Chronicle*, September 2, 2005. http://www.chron.com/news/hurricanes/article/The-foretelling-of-a-deadly-disaster-in-New-1923023.php#page-1 (accessed 12 December 2013).

Blake, Eric S., Christopher W. Landsea and Ethan J. Gibney (2011). "Retelling-of-a-deostliest, and most intense United States tropical cyclones from 1851 to 2010 (and other frequently requested hurricane facts)." http://www.nhc.noaa.gov/dcmi.shtml (accessed 12 December 2013).

Bourne, Joel K. Jr. "Gone with the Water." *National Geographic Magazine*, October 2004. http://ngm.nationalgeographic.com/ngm/0410/feature5/index.html (accessed 12 December 2013).

Fischetti, Mark. "Drowning New Orleans." *Scientific American*, September 5, 2008. http://www.scientificamerican.com/article.cfm?id=drowning-new-orleans-hurricane-prediction (accessed 12 December 2013).

Goldenberg, Suzanne. "US urged to abandon ageing flood defences in favour of Dutch system." *The Guardian*, 5 June 2009. http://www.guardian.co.uk/environment/2009/jun/05/flooding-us-defence (accessed 12 December 2013).

Morris, Ashley. "Fuck you, you fucking fucks." *Ashley Morris: the blog*, 27 November 2005. http://ashleymorris.typepad.com/ashley_morris_the_blog/2005/11/fuck_you_you_fu.html (accessed 12 December 2013).

The Times-Picayune. "Transcript of President Barack Obama's Katrina Anniversary Speech at Xavier University." August 29, 2010. http://www.nola.com/katrina/index.ssf/2010/08/transcript_of_president_barack.html (accessed 12 December 2013).

Daniela Agostinho
Flooded with Memories: Risk Cultures, the Big Flood of 1953 and the Visual Resonance of World War Two

> Events are the foam of things, but what I am interested in is the sea.
> Paul Valéry

In the introduction to the volume *Natural Disasters, Cultural Responses*, Christof Mauch argues that the memory of natural disasters, in contrast to the memory of war, is markedly short-lived: "As waters return to their preflood levels and the last victims of earthquakes are recovered, so, too does mass-media interest in natural disasters quickly subside" (2009, 3). This is certainly not the case with the Netherlands, where the occurrence of floods was and is as decisive for identity formation as any past political and military disputes. In a short story symptomatically entitled "The Netherlands Lives with Water", published in the American literary magazine *McSweeney's*, Jim Shepard imagines a Dutch hydraulic engineer who works on managing rising sea levels and keeping what the old people called the "Waterwolf" from the door. Shepard wrote the story after being asked to contribute to an issue of *McSweeney's* in which all stories were set some decades in the future. Interested in climate change, he decided to place his fiction in an endangered Rotterdam, which he said was "dumpier" than Amsterdam and The Hague but "more economically crucial to the country and in much more immediate danger". "It's the catastrophe for which the Dutch have been planning for fifty years" (Shepard 2009, 191), the protagonist explains (192):

> Or, really, for as long as we've existed. We had cooperative water management before we had a state. The one created the other; either we pulled together as a collective or got swept away as individuals. The real old-timers had a saying for when things fucked up: "Well, the Netherlands lives with water." What they meant was that their land flooded twice a day.

Shepard's text echoes Simon Schama's definition of the Netherlands as a "hydrographic culture" (Schama 1988, 44), a society where state structures originated from systems of local responsibility for dike and dam maintenance (cf. Eßer 1997). However, one may also take this notion of "hydro-graphic" as the inscription of the struggle against water in the cultural fabric through narratives and representations. Shepard's hydrographic story also falls within Scott Lash's notion of "risk culture", which revises Ulrich Beck's concept of risk society. Risk culture terms a community where risk is perceived not so much through instrumental rationality or normative thinking but "through symbolic practices and specially through symbol innovation" (Lash 2000, 60). Lash criticizes the remaining strength of institutionalist norms in Beck's concept of risk society, claiming there are areas of liminality, indeterminacy and subjectivity that Beck's notion does not account for. While the notion of risk society would always presume "a determinate, institutional, normative, rule bound and necessarily hierarchical ordering of individual members in regard to their utilitarian interests" (47), the concept of risk culture comprises all kinds of residual forms of sense-making practices. Whereas Beck points to the incalculability of techno-hazards while at the same time seeking solutions in the conventional realm of science and politics, Lash emphasizes the need to mobilize resources that were hitherto considered personal, subjective, intuitive and non-specialized.[1]

Drawing on Kant's distinction between determinate and aesthetic judgment – the former associated with closed, prepositional truths, the latter with open estimations based on bodily feelings and the imagination –, Lash considers the implications of aesthetic reflexivity as a foundation for socio-political action. Even though, Lash stresses, all aesthetic judgments bridge perception and reason, reflexive or "aesthetic judgments are estimations that are based on 'feelings' of pleasure and displeasure, but also on feelings of shock, overwhelmedness, fear, loathing as well as joy" (53). Risk cultures are therefore "reflexive communities" (47), reflexivity being understood as an active rather than passive response to threatening conditions. Unlike risk society, the governing figures of risk culture are not rules but symbols. Deborah Lupton, who also refers to aesthetic judgments as "hermeneutical reflexivity", argues that this "involves the processing of signs and symbols rather than simply 'information'" (Lupton 2005, 121). Cultural representations are thus mediators through which communities not only reflexively judge risk events but also shape and refashion the

---

1 Lash also reviews Mary Douglas' and Aaron Wildavsky's position in their work *Risk and Culture* (1983), which he considers "hierarchical institutionalist" (Lash 2000, 55) as it privileges normatively ordered institutions to the detriment of what they term, in a conservative vein, "sects" or "margins of society". Douglas and Wildavsky argue that there is in fact no increase in risks in contemporary societies but rather an increase in perceived risks that are constructed by the groups "from the borders". Lash contrasts this conservative position to the respective notions of risk society posited by Beck and Giddens, which despite being ultimately institutionalist are not concerned "with social control of radicals, but with a measure of amelioration of environmental and identity-risks" (Lash 2000, 49). Their argument, quite different from that of Douglas and Wildavsky, involves modern institutions in fact giving rise to a number of hazards whilst simultaneously attempting to handle them. In so doing, they have been partly successful but have also produced the side effects of still further risks.

sensibilities of their risk cultures. Our relation to contemporary environmental hazards, Lash contends, is not and cannot be grounded, or at least not exclusively, in any determinate judgment; instead, it is inescapably tied to the world of imagination and emotions. Lash cites the example of the legendary Robert Mapplethorpe exhibition that displayed violent homoerotic photos at several art museums throughout the world between 1990 and 1995. Mapplethorpe's own death from AIDS crossed the mind of many viewers of the exhibition while the issue of AIDS in general definitely stood out as a pressing concern. This is a different process of judging the risk of AIDS than assessing it through statistics or scientifically based knowledge. As Lupton contends, "aesthetic and hermeneutic reflexivity is not rooted in self-monitoring, but rather in self-interpretation, involving intuition and the imagination above moral and cognitive judgment" (2005, 121). It emerges out of interaction with a particular cultural artifact, a symbolic representation, which can potentially open up a space of reflexive and subjective risk awareness. To understand risk in terms of statistical monitoring is important as far as public policy is concerned, but this may not be sufficient since the way people perceive, experience and represent risk should also be a matter for policy-making. The victim experiences of risk are about more than just policy. They also incorporate systems of meaning, situated knowledges and beliefs, shared experiences, imagination, identity and uncertainties. In this sense, all cultures need artifacts, representations, narratives through which people can recognize risk and assess future threats from a subjective, aesthetic and affectively-charged point of view.

Symptomatically, in an article in the German weekly paper *Die Zeit*, entitled "The Power of Images", Florian Illies claimed that because the world saw the images of the Fukushima disaster, faith in the human capacity to dominate technology has collapsed (Illies 2011). The article goes as far as to say that Fukushima is the second image of the twenty-first century after the falling towers of 9/11, doomed to join, to quote W. J. T. Mitchell, "the iconic mushroom cloud as the principal emblem of terror in our time" (2011, 78). Both images – 9/11 and Fukushima – have had the effect of contributing to a new state of insecurity that sets the tone for twenty-first century risk culture but also of fostering new formations of power, ideology and social critique. Visual representations of natural disasters from around the world have been generating the feeling that catastrophes are becoming more frequent, more threatening and more devastating. As Brian Massumi has pointed out, individuals in late modernity experience a constant fear, "a kind of background radiation saturating existence" (1993, 24). And the fact that he employs the term "radiation" is anything but innocent. If it is true that images – or cultural representations at large – are not the exclusive agents of the "sensibilities of risk culture" (Lash 2000), it is also undeniable that they do have an effect on knowledge, sensibilities and affects that cannot be overlooked. Indeed, as Horst Bredekamp (2010) has argued, drawing from speech act theory, images are effectual, they produce as much as they reproduce; they are "image acts" that create facts by bringing images into the world and thereby shape our knowledge, perceptions and anxieties.

Throughout history, the populations of the Dutch coastal provinces have been regularly afflicted by devastating storm surges. Most famous are the St. Elisabeth Flood of 1421 and the All Saints' Day flood of 1570, which cost the lives of many thousands of people and caused enormous damage. The Great Storm of 1703 and the Flood Disaster of February 1825 also had quite a destructive effect. These structuring events, amongst many others, shaped the risk matrix against which our case study unfolded: the so called Big Flood of 1953, the *Watersnood-ramp*, a disaster that hit the South-west of the Netherlands on the night of Saturday 31 January and the morning of 1 February. It affected mostly the Dutch provinces of Zeeland, South Holland and Noord-Brabant, taking the lives of 1,836 people. Considered one of the most destructive floods of the past century in Europe, it is now retrospectively referred to as the "Dutch Katrina", proving that disasters become templates that cross temporalities and contaminate each other's representations. This disaster may also be regarded as the catalyst of a new paradigm of disaster preparedness since it led to the construction of the Delta Works, an elaborate and innovative project involving the closing off of most estuary-mouths, regarded as a milestone in engineering and a role model in water management.[2]

---

2 Instead of analyzing past floods and building sufficient protection to deal with them, the Delta Works commission pioneered a conceptual framework to use as norm for investment in flood defenses, which became the new paradigm in water management and flood prevention. Within this new model, the chances of a significant flood within a given area are calculated, using data from a purpose-built flood simulation lab, empirical statistical data regarding water wave properties and distribution. Storm behavior and spring tide distribution were also taken into account. A number of acceptable risks were contemplated in the Delta law, requiring the government to keep the risks of catastrophic flooding within those limits and to upgrade defenses should new insights into risks demand such actions. Due to climate change and relative sea-level rise, the dikes will eventually have to be made higher and wider. In a report in September 2008, the Delta commission advised that the Netherlands would need a massive new building program to strengthen the country's water defenses against the anticipated effects of global warming for the next 190 years. This commission was created in September 2007 after the damage caused by Hurricane Katrina to New Orleans prompted a wider reflection on such problems.

In what follows, however, I argue that the national history of disasters was not the only framework that shaped the reception of this major flood. The fact that it occurred eight years after the end of World War II has not yet been appropriately taken into account. The traumatic events of WWII, and especially the bombardments of The Hague and Rotterdam in 1940 and 1945, provided a second backdrop that formed the perception and response to this calamity. In fact, in the case of the Big Flood of 1953, the intertwinement of memories of war and of natural disasters contributed, on the one hand, to the selective remembrance of the war experience and, on the other, to a new degree of risk awareness. This paper thus contends that the singularity of the Big Flood of 1953 is contested by the interplay between memories of war and memories of disaster that resulted in a non-singular representation of the disastrous event. To demonstrate this interplay, I examine the photographs of the Big Flood of 1953 taken by Ed van Wijk, who during Second World War had photographed the bombardments of Rotterdam and The Hague. I wish to claim that the visual memory of WWII functioned as a cultural template for the representation and reception of the Big Flood and that this cultural transference was in part responsible for a new type of reflexive risk judgment.

1 Water, War and the Non-singularity of Disaster

In Jim Shepard's hydrographic story, a future where the safety provided by the Delta project is being threatened by climate change and rising waters is set against a historical past that shapes the narrative present and the imagined future. Shepard's short story is exemplary in intertwining the historical layers that form the cultural fabric upon which Dutch risk culture gained form. First of all, the long history of facing and dealing with disaster that has so profoundly carved the country's self-image:

> We'll come up with something. We always have. Where would New Orleans or the Mekong delta be without Dutch hydraulics and Dutch water management? And where would the U.S. and Europe be if we hadn't led the way out of the financial Panic and depression, just by being ourselves? E.U. dominoes from Iceland to Ireland to Italy came down around our ears but there we sat, having been protected by our own dutchness. What was the joke about us, after all? That we didn't go to the banks to take money out; we went to put money in. [...] Who tells anyone who'll listen that we're providing the rest of the world with a glimpse of what the future will be? (Shepard 2009, 196).

The country's proud competence in water management is thus extended to other domains of action and problem solving. The current financial crisis in

Europe is one of the many troubles that the Dutch, profiting from long-established know-how, have managed, in Jim Shepard's story, to overcome and thereby validate their "own dutchness". This reflexive notion of all-encompassing efficiency derives from the formation of what Greg Bankoff (2003) has termed a "culture of disaster", a set of rehearsed practices of risk minimization and strategies to cope with extraordinary situations that have been integrated into daily lives to avoid being taken by surprise.[3] Cultures of disaster are also, therefore, "cultures of coping" (Bankoff 2009), of practices that differ according to each social and cultural milieu,[4] sustaining a feeling of safety and resilience based on risk expectations and cumulative knowledge. As Franz Mauelshagen has argued, "communities' coping strategies must be seen as the result of a series of occurrences that are perceived as similar and recurring even though their unpredictable appearance still defines them as discrete events" (2009, 44). This implies that disasters are not singular but rather recurring and comparable. Indeed, "cultures of coping" depend on the accumulation of knowledge from repeated disasters. As such, Mauelshagen further explains, "disasters are neither unique nor exceptional from a social or historical perspective, even though they may be from a personal or generational one" (43). If they were unique and exceptional, no cumulative knowledge could be formed.

In fact, Jim Shepard's fiction concludes with a future Rotterdam struck by a massive storm tide while the protagonist evokes his mother's memories of the 1953 flood, which emerges throughout the story as a template, as a seminal event against which the future catastrophe is gauged:

Cato [the protagonist's wife] was moved by all of my mother's flood memories, but was only brought to tears by one. My mother's only cherished memory from that year: *the Queen's address to the nation afterward, and her celebration of what the crucible of the disaster had produced: the return, at long last, of that unity the country had displayed during the war.* My mother had purchased a copy of the speech on LP, all those years ago and had had her neighbor transfer it to a digital format. She played it for us [...] while the Queen's smooth and warm voice thanked us all for the way we had worked together in that one great cause, soldiering on without a thought for care, or grief, or inner divisions, and without even realizing what we were denying ourselves (Shepard 2009, 212; emphasis mine).

---

3 Greg Bankoff (2003, 2009) applies this notion to his study of society and natural hazards in the Philippines.
4 Cf. the text of Zaumseil and Prawitasari-Hadiyono in this volume, examining the "culture of coping" in Indonesia, a country that despite being a former Dutch colony displays a coping system entirely different from the instrumental rationality of the Dutch control paradigm. It comprises a different conception of preparedness, more spiritual than instrumental, and a readiness to accept disaster instead of attempting to control it. Similar to the Netherlands, however, there is also a sense of pride in the local or national "culture of coping".

What Shepard's story suggests, in weaving the experience of natural catastrophes with economic crises and memories of war, is that disasters, regardless of their nature, are not singular events, even if they always entail a unique character from the personal point of view. Catastrophes always unfold against a cultural palimpsest of previous disasters that frames their reception, endows them with meaning and shapes people's responses.

In Shepard's imagination of disaster, in his anticipation of future threats, the memories of the 1953 Flood and of the Second World War function as narrative blueprints. Both can be regarded as "impact events", which Anne Fuchs, referring to the Dresden bombardments, defined as "historical occurrences that are perceived to spectacularly shatter the material and symbolic worlds that we inhabit" (Fuchs 2010, 37). The notion of "impact" is critical here in that it suggests the afterlife of catastrophic events in the material culture and collective consciousness. Impact events often reverberate during the unfolding of new disasters, their "hidden legacy" (38) rising to the surface of a new context that reactivates their memory. Andreas Huyssen, for instance, has shown that the memory of the Allied bombings over German cities resurfaced in German society at the beginning of the Iraq war in 2003 when the German peace movement strategically invoked the bombardments as a template of suffering and victimhood to contest the military intervention in Iraq. According to Huyssen, when German protesters marched against the new Gulf war under the slogan "We know what it's like to be bombed", the past re-emerged as "citation, as selective recall with a claim to absolute presentness – generational difference of experience eliminated, a collapsing of time zones into a broadened present" (2003, 165).

World War II – and the aerial bombings in particular – is one of those "impact events" that resurface in later historical moments to help the affected communities make sense of different disastrous occurrences.[5] Dutch historian Krijn Thijs has contended that the memory of WWII and the Nazi occupation of the Netherlands often emerge as a template in the historical present of Holland, functioning as "orientation and reference", as a "moral anchor and benchmark for comparison", often more latently than explicitly (2011, § 2).[6] In fact, WWII in

---

5 Silke Arnold de Simine (2007) also argued that the memory of the Blitzkrieg proved to function as a template for the 7/7 bombings in London. Although most Londoners of today did not experience WWII or the Blitz, these events are nevertheless part of cultural memory and the connections between the terror bombings and the German "terror raids" were easily made.
6 Translation mine. Complete original version: "Die Besatzungszeit 1940–1945 bildet nach wie vor das Rückgrat der niederländischen Erinnerungskultur. Sie dient der unsicheren Gegenwart als Orientierung und Referenz, auch in jüngeren politischen Debatten. In Kontroversen über

the Netherlands is also known as simply "the war" (*de oorlog*), while the Big Flood of 1953 is referred to as "the disaster" (*de ramp*). This rhetorical singularization lays bare the profound impact of these two events in society and culture whereby they became watershed moments, caesuras in historical time, ruptures of the normal order. At the same time, however, the insistence on the singular beclouds historical continuities, cultural patterns and temporal exchanges that are decisive for the way people perceive, experience, respond and process those events. Indeed, in spite of their historical contingency and particular nature – natural, technological or manmade – impact events merge at the levels of experience and representation, cutting across and binding together different temporalities. Michael Rothberg terms this process "multidirectional memory", which he defines as the "dynamic transfers that take place between diverse places and times during the act of remembrance" (2009, 11) and which enable moments of transaction between different impact events.[7] Moreover, impact events rely on "impact narratives" (Fuchs 2010) to act out their influence on the symbolic order. Impact narratives convey and at the same time constrain the impact of catastrophes through images, stories and artifacts that articulate traumatic experiences and conform the memory of past disasters to the situational demands of the present. Through analysis of Ed van Wijk's photographs, the next section examines how war memories shaped the representation of the Big Flood of 1953 as a cultural narrative summoned to help overcoming the disaster of the present.

---

die multikulturelle Gesellschaft und ihr vermeintliches Scheitern, über die Gleichberechtigung verschiedener Religionen im öffentlichen Raum, über den Erfolg des Rechtspopulismus und seiner umstrittenen 'Freiheitsverteidiger' – bei allen diesen Themen dient die Besatzungszeit, wenn auch häufiger latent als explizit, als moralischer Anker, als Vergleichsmaßstab und letztgültiges Argument."

7 Rothberg developed this theory having the Holocaust as its main reference to argue for a more transnational and transcultural approach to the event in order to tackle the relationship between the different histories of victimization of social groups and enable the historical recognition of other violent events, such as colonialism and other genocides, that have been overshadowed by the uniqueness of the Holocaust. His proposed multidirectional model of memory attempts to displace a model of "competitive memory" that regards different histories of violence as competing with each other for recognition; instead, he appeals to a model of negotiation and interaction of multiple historical memories, in which Holocaust consciousness, for instance, can work as a platform to articulate a vision of other traumatic pasts.

2 Impact Narratives and Visual Resonances

Ed van Wijk (1917–1992) was among the many prominent Dutch photographers, like Ed van der Elsken and Aart Klein, who covered the Big Flood of 1953.[8] During the war he photographed the destruction of Rotterdam, the evacuation of the Marlot quarter in The Hague and the village of Wassenaar, the bombardment of Bezuidenhout also in The Hague and, finally, the Liberation. His work is therefore a valuable resource to examine the impact narratives of these two seminal events and their contribution to a reflexive or hermeneutical risk culture.

In the introduction to his book *De Ramp* [The Disaster], Kees Slager writes: "In the hours during which the disaster took place, no photographs were made at all; it was dark and besides, people had other things on their minds. Most photographs of the disaster were taken after the storm had died down, and the water was therefore a lot less wild" (2003, 7). Against this backdrop of immediate aftermath, Ed van Wijk's images of the Big Flood convey the unexpectedness and enormous devastation of the disaster whilst capturing the ensuing relief effort and the spirit of bouncing back.

The far-reaching devastation of the disaster is pictured through wide shots of flooded landscapes, which testify to the defeat of man by a forceful nature he had tried to transgress and master. In these pictures, the water always overwhelms the soil, with the seashore being off-framed and overpowered into invisibility (fig. 1) as an attempt to represent the destructiveness of a disaster that was not possible to capture when it struck.

---

8 Dolf Kruger and Kees Molkenboer were also among the photographers who covered the Flood. Not only are the different approaches of the various photographers evident but also the influence of the patrons, namely a press that was strongly compartmentalized along sociopolitical lines. Dolf Kruger worked for the communist newspaper *De Waarheid*, Kees Molkenboer's photographs appeared in the "neutral" *Rotterdamsch Nieuwsblad*, Aart Klein published in the Christian journal *De Spiegel* and Ed van Wijk in the liberal newspaper *Het Vaderland*.

Fig. 1: The Big Flood of 1953. © Ed van Wijk/Nederlands Fotomuseum

Many of these pictures seem to refigure and remediate previous representations of flood disasters in the Netherlands, as if harking back to prior events to face and process the present catastrophe (fig. 2). The February flood of 1825, the most devastating flood disaster before 1953, emerged as a template of national "disaster culture" to be actualized in new representations. The February flood was the worst natural disaster of the nineteenth century but was quickly forgotten once the damage was repaired and no major political or engineering repercussions ensued. The disaster of 1953, however, retrieved it from oblivion and reactivated its representational formulae in a new context.

Fig. 2: February Flood 1825. Drawing D. de Hoop

"Remediation" (Bolter and Grusin 2000), the process through which new media refashion older media, appropriating their "techniques, forms, and social significance" (65), is crucial in this process of retrieval. On the one hand, the photographs of 1953 reemploy the tropes of paintings and drawings of the disaster of 1825, which provide the present with an impact narrative to work with and build upon, whereby the singularity of the events is challenged on the representational level. On the other hand, photography, despite remediating previous media, proves a medium with specific particularities both subtracting and adding new dimensions to previous representations of disaster, such as its indexical relation to the reality depicted. While paintings and drawings have a tradition of portraying disasters as they unfold, the photographic coverage of the Big Flood had to adjust to the contingencies of the present and focus on the aftermath. As such, remediation both disputes singularity and generates new formulae and meanings, contributing to the formation of what Mauch termed "cumulative disaster knowledge", the acquisition of knowledge based on previous occurrences in which new incidents always constitute an addition to the prior understanding of catastrophe.

Fig. 3: The Big Flood of 1953. © Ed van Wijk/Nederlands Fotomuseum

The pictures of devastation contrast with those of the disaster relief efforts (fig. 3), translating a timeline of struggle between the crushing nature and a community caught unaware but willing to face up to the challenge after the shock as standard "disaster culture" behavior. A visual narrative of communal effort in the disaster relief and damage control was clearly preferred to the detriment of the portrayal of major losses, assistance failures and unpreparedness. Once again, the past provided a strategic model of representation that proved to fit the demands of the present. Ed van Wijks photographs of the relief efforts seem to remediate another depiction of the 1825 flood (fig. 4), reemploying the tropes of the windmill (a symbol of the hydrographic identity to which the photographs tried to appeal) and the resourceful men on the boat trying to retrieve people from inside the house. Moreover, by framing the destroyed houses together with the shore, the photograph and the drawing translate a sense of grounding and bouncing back, as opposed to the images in which the seaside was deliberately kept out of view to convey a stronger impression of devastation that ends up accentuating and praising the heroism and resilience of the civilians involved in the disaster relief.

Fig. 4: *Watersnood te Buiksloot* (J. Claus jr., 1825)

The bird's eye view is one of the most recurring features of disaster representation and it was greatly deployed in depictions of the Big Flood. If, on the one hand, it renders the immense extent of the flood, emphasizing the human effort to overcome it, on the other hand, it conveys a safe viewer's perspective, a secure standpoint to contemplate the catastrophe from, as figure 1 attests. In *Shipwreck with Spectator*, Hans Blumenberg discusses the Lucretian spectator who watches the distress of those at sea from the safety of dry land, who experiences pleasure not as a result of seeing someone else suffer but of enjoying one's own safe standpoint: "By virtue of his capacity for distance from the disaster, he stands unimperiled on the solid ground of the shore, rejoicing over his own self-preservation" (1997, 17). The bird's eye view in this case can be regarded as meta-representational for it demonstrates a further safe perspective, that afforded by representation, which despite conveying the impact of catastrophe and further exerting it onto society's risk culture also contains the impact inside the frame and the temporality of representation, guaranteeing safety at the standpoint of reception.

The same photographer had adopted a similar perspective in works shot in Rotterdam in 1940, prior to Nazi occupation, in which the contrast of the wide angle and the small human figures renders the immensity of the destruction and human desolation, with the wreckage extending beyond the horizon. Through this gaze from above, similar to that of Wim Wenders's angels in *Der Himmel über Berlin*, the distance from the destruction conveys an almost detached, trans-historical perspective that assures safety upon looking.

Fig. 5: After a bombardment in Rotterdam, 1940. © Ed van Wijk/Nederlands Fotomuseum

This "singularity of a form in perpetual variation" (Deleuze 2005, xiv) may be grasped under the notion of resonance introduced by Aleida Assmann, which she defines as the "evoking or suggesting of images, memories, emotions and meanings" (2010, 6) in a new context. A relation of resonance presupposes, according to Assmann, the interaction of two different elements, one situated in the foreground and the other in the background: "The element in the foreground does not cover up or elide that existing in the background; on the contrary, the element in the foreground triggers the background and fuses with it. We may also speak of cooperation, in which the background element non-consciously or un-consciously guides, forms and shapes the foreground element" (6).

This would explain how the representation of WWII bombardments prefigured and shaped the depiction of the Big Flood of 1953 or, from the opposite perspective, how the Big Flood triggered or alluded to the visual memory of the bombings, which emerges as a cultural template.[9] Indeed, images are the outcome of past and present impressions that conjoin to generate new meanings. Images might be said to produce or suggest worlds hitherto unseen but always emerging from the already seen. As cultural theorist Thomas Macho (2010) has claimed, images can relate both to the past and to the future. In his book *Vorbilder* ("paragons" but literally "pre-images"), he proposes that these paragons, these prior images are also often future-images (*Zukunftsbilder*): like archetypes or blueprints, they envision what is yet to exist. They both remember and anticipate, evoke and invoke. This dynamic, trans-historical transfer is charged with political significance. As seen before in Jim Shepard's short story, the evocation of WWII in the remembrance of the Big Flood was vested with a positive message of national union and the spirit of endurance when the mother of the protagonist recalled the uplifting address of the Queen to the nation after the flood, appealing to the "unity the country had displayed during the war" (Shepard 2009, 212). Indeed, the fact that the flood took place only eight years after the war prompted comparisons and exchanges. In a news article published the day after the flood, a citizen even described the impact of the disaster as such: "It was like a city after a bombardment."[10] Symptomatically, images of the flood often portrayed destroyed houses that resembled bombarded buildings (fig. 3).

As argued earlier, impact events are inseparable from the ways in which they are received and interpreted by individuals and communities which process them according to changing social and cultural demands. One of the effects of visual resonance relations lies in providing the present in distress with a disastrous image of the past that can function as a steering example of how to overcome the catastrophic occurrence. In the photographs of the Big Flood, the visual resonance of the bombardments comforts and encourages people, pro-

---

9 Assmann considers the concept of resonance a variant of reception theory as it implies the remembrance of images and stories from the point of view of the receiver and the "interaction between an input on the surface – be it an experience, a text, an image or a melody – and the deep structure of internalized cultural templates and schemata" (Assmann 2010, 16).

10 "Als een stad na een bombardement", *Nieuwsblad van het Noorden*, 2 February 1953, p. 8.

moting their sense of community and urging them to react in the same way as before. This selective memory of the war could be seen as reassuring as it reminded people that something more awful had already occurred in the past and nevertheless people had been able to make it through. As Aleida Assmann puts it: "Wherever an impact event is answered by an accepted impact narrative, this narrative transforms the event and eventually replaces it. In this process, the narrative absorbs the hot kernel of the event and appropriates it in a new framework of religious or political meaning" (2010, 9). Indeed, the master-narrative of the war during the 1950s privileged a discourse of national cohesion over victimhood (cf. van Vree 1995) that shaped the response to the flood. This narrative of national unity was thus strategically invoked to foster acts of solidarity and new communal identities, while both the traumatic experiences of war and the flood were deliberately off-framed through carefully selected tropes. As such, a complex process of displacement resembling a Freudian screen memory was set in motion. A screen memory was defined by Freud (1899) as an image that stands in for either a prior or a subsequent disturbing memory. It disguises the original shocking memory while the persisting vividness of the screen memory simultaneously denounces the impact of that experience in the psychic world. In a similar process, a selective memory of the war, on the one hand, screened out the negative effects of the flood to focus on the spirit of unity and reconstruction; on the other, the emphasis on the relief effort in the images of the Flood contributed to gloss over the traumatic images of destruction of 1940 and 1944. In Freud's terminology, an early memory was used as a screen for a later event, which, in turn, screened back the early event. Instead of providing a platform to act out and articulate concealed and painful memories, the merging of images of WWII with those of the Big Flood activated a rather selective remembrance to foster national sentiments and a sense of endurability, while eliding traumatic experiences and generating a newly accepted narrative that stands for the event itself: a watershed moment that prompted disaster preparedness.

A second effect of visual resonance relations involves contributing to a growing matrix of disasters that prove that catastrophes are recurring and that one has to remain alert. Indeed, one of the reasons that explains the unpreparedness for the Big Flood of 1953 is the so called "disaster gap", the rare non-occurrence of severe hazards throughout almost a century, which can explain the loss of "functional disaster memory" (Pfister 2011, 17). Hence the remediation of the February flood of 1825 – the last major disaster – and the visual resonance of WWII, the last impact event with major consequences for society and material culture. Both modes of retrieval can be seen as a strategy to keep the memory of past disasters alive for the future risk culture and to encourage a stronger awareness of risk and the political commitment of the public.

Fig. 6: The Big Flood of 1953. © Ed van Wijk/Nederlands Fotomuseum

The inscription of a spectatorial position within these images is decisive to attaining this purpose and conveying the desired viewpoint and interpretation of the events. In addition and in contrast to the safe perspective of the bird's eye view, Ed van Wijk's images of the Big Flood also placed viewing subjects within the debris, fostering an at risk subjectivity. Indeed, van Wijk's images convey a progression from the safety of distance to a growing embedment in the wreckage of disaster (cf. figs. 1–6).

The protective distance from disaster collapses and this new spectator is exposed to the impending dangers on the horizon, forced to acknowledge and prepare for future and not yet visible risks. Moreover, the spectatorship is configured collectively, emphasizing the need for an expansive communal disaster preparedness. This collective witnessing of present and impending danger was also prefigured in WWII (fig. 7), in the depiction of the bombardments in The Hague, where the subjects, no longer in safety, observe destruction from within the debris and realize their increasing exposure to danger and insecurity, which would become the master-narrative of the ensuing cold war.

Fig. 7: After a bombardment in The Hague, 1944. © Ed van Wijk/Nederlands Fotomuseum

However, as in David Hockney's *Picture Emphasizing Stillness*, in which the painter, commenting on the safety ensured by the image's stillness, inscribed the sentence "They are perfectly safe – this is a still" between two humans and a wild animal, the pictures of the flood and the bombardments, despite inscribing the subjects in the midst of disaster, are also still images that protect the viewers outside the picture from harm by containing the danger inside the frames of representation.

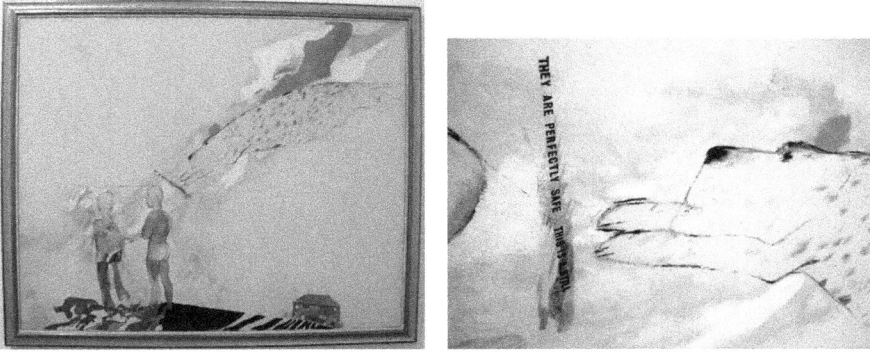

Fig. 8: David Hockney, *Picture Emphasizing Stillness,* 1962. Berardo Collection, Lisbon

At the same time, images of disaster also contain the potential to disquiet and appeal to the vulnerability of their viewers. One last image of a bombardment in Rotterdam proves exemplary (fig. 9). It pictures a group of onlookers staring in two directions. One backwards, and the others forward. The spectators looking forward, the main focus of the picture, face the present catastrophe and try to foresee the invisible future risks hovering over their community. As one of Jim Shepard's characters prophesizes: "Man had created war and the Bomb and now nature was going to exact its revenge."

The spectator looking backwards, to the off-frame of the picture, searches the debris of the past for a meaning for the present and a prefiguration of the future. Just like the spectator of Lucretius, who, according to Blumenberg, "no longer needs a sailor in peril at sea, because he projects his own past or future distress into the image of the raging sea" (1997, 65). However, this spectator is also facing us, the viewers of the pictures, and by looking outwards, he inscribes our own gaze in the picture, producing a double spectatorship of risk. On the other hand, by attempting to bridge the gap between the space of representation and that of reception, he inscribes the afterlife of disaster inside its representation. Through his outward-looking gaze, he is confronting the future viewers of this image, asking them to abandon their standpoint of safety and acknowledge the impending and invisible hazards on the horizon, as if warning them that the danger is not locked inside the stillness of representation, but right outside in the empirical world. Not surprisingly, the same shot was replicated in various pictures of the Big Flood with destructed houses in the background, resembling, once again, the bombarded buildings of Rotterdam and The Hague, which had served as a blueprint.

Fig. 9: After a bombardment in Rotterdam, 1940. © Ed van Wijk/Nederlands Fotomuseum

As Zygmunt Bauman has argued in *Postmodernity and its Discontents*, "the growing awareness of the dangers ahead goes hand in hand with a growing impotence in preventing them or alleviating the gravity of their impact" (1997, 55). The interpellating look of the spectator inside the picture at us reminds us of the modern ambivalence between disaster preparedness and unpredictability, inciting us to remain vigilant, working therefore as a "premediating" gaze. Differently to prefiguration – the existence of templates, which act as blueprints for the construction of images and the patterns of their reception (which, as discussed earlier, also applies here) – premediation (Grusin 2004) works through the pluralization of scenarios in which the future can be anticipated and controlled in more subtle ways. Grusin's primary reason to introduce the concept of "premediation" was to discuss the role of the mass media in coping with terror. From his perspective, the ultimate goal of premediation is the "colonization of the future": "premediation seeks to make sure that the future is so fully mediated by new media forms that it is unable to emerge into the present without having already been remediated in the past" (2004, 36). Premediation in this sense means discussion of future scenarios, narratives of the future and discourses on risk that are designed to deprive the future of its undesired shock quality. As Susan Sontag suggested in her essay about the "Imagination of Disaster", future scenarios can help us "cope with spectres" and "normalize" what is unbearable to imagine, "thereby inuring us to it" (Sontag 1965, 42). The process of visual displacement that occurred in the representation of the Big Flood of 1953, in which WWII emerged as a resonating blueprint, can thus be grasped as the premediating impact narrative that attempted to foster risk awareness and disaster preparedness in contemporary Dutch risk culture. As Schiller once said, as if recognizing the role of visual premediation for the anticipation of catastrophe, "our safety is not in blindness, but in facing our danger" (2006, 109).

References

Bankoff, Greg. Cultures of Disaster. Society and Natural Hazards in the Philippines. London: Routledge, 2003.

Bankoff, Greg. "Cultures of Disaster, Cultures of Coping: Hazard as a Frequent Life Experience in the Philippines." In Christof Mauch and Christian Pfister (eds.), Natural Disasters, Cultural Responses. Case Studies Toward a Global Environmental History, 265–284. Lanham, MD: Lexington Books, 2009.

Bauman, Zygmunt. Postmodernity and its Discontents. Cambridge: Polity Press, 1997.

Beck, Ulrich. World at Risk.Cambridge: Polity Press, 2009.

Blumenberg, Hans. Shipwreck with Spectator: Paradigm of a Metaphor for Existence. Trans. Steven Rendall. Cambridge, MA: MIT Press, 1997.

Bolter, Jay David, and Richard Grusin. Remediation. Understanding New Media. Cambridge, MA: MIT Press, 2000.

Bredekamp, Horst. Theorie des Bildakts. Frankfurter Adorno-Vorlesungen 2007. Berlin: Suhrkamp, 2010.

Deleuze, Gilles. Francis Bacon: The Logic of Sensation. Trans. Daniel W. Smith. London: Continuum, 2005.

Douglas, Mary, and Aaron Wildavsky, eds. Risk and Culture. Berkeley: University of California Press, 1983.

Eßer, Raingard. "Fear of Water and Floods in the Low Countries." In William G. Naphy and Penny Roberts (eds.), Fear in Early Modern Society, 62–77. Manchester: Manchester University Press, 1997.

Freud, Sigmund. "Screen memories." In Standard Edition, vol. 3, 301–322. London: The Hogarth Press, 2001 (1899).

Fuchs, Anne. "The Bombing of Dresden and the Idea of Cultural Impact." In Rebecca Braun and Lyn Marven (eds.), Cultural Impact in the German Context. Studies in Transmission, Reception, and Influence, 36–57. New York: Camden House, 2010.

Grusin, Richard. "Premediation." Criticism 46, no. 1 (2004): 17–39.

Huyssen, Andreas. "Air War Legacies: From Dresden to Baghdad." New German Critique 90 (2003): 163–176.

Lash, Scott. "Risk culture." In Barbara Adam, Ulrich Beck and Joost van Loon (eds.), The Risk Society and Beyond: Critical Issues for Social Theory, 47–62. London: Sage, 2000.

Lupton, Deborah. Risk. London: Routledge, 2005.

Macho, Thomas. Vorbilder. Munich: Wilhelm Fink, 2010.

Massumi, Brian, ed. The Politics of Everyday Fear. Minneapolis: University of Minnesota Press, 1993.

Mauch, Christof. "Introduction." In Christof Mauch and Christian Pfister, eds. Natural Disasters, Cultural Responses: Case Studies toward a Global Environmental History, 1–16. Lanham: Lexington Books, 2009.

Mauelshagen, Franz. "Disaster and Political Culture in Germany since 1500." In Christof Mauch and Christian Pfister, eds. Natural Disasters, Cultural Responses: Case Studies toward a Global Environmental History, 41–75. Lanham: Lexington Books, 2009.

Mitchell, W. J. T. Cloning Terror: The War of Images, 9/11 to the Present. Chicago: University of Chicago Press, 2011.

Rothberg, Michael. Multidirectional Memory. Remembering the Holocaust in the Age of Decolonization. Stanford: Stanford University Press, 2009.

Schama, Simon. The Embarassment of Riches. An Interpretation of Dutch Culture in the Golden Age. Berkeley and Los Angeles: University of California Press, 1988.

Schiller, Friedrich. Aesthetical and Philosophical Essays. Middlesex: The Echo Library, 2006.

Shepard, Jim. "The Netherlands Lives With Water." McSweeney's Quarterly Concern 32 (2009): 189–212.

Simine, Silke Arnold-de. "London can take it: remembering the air war. Reliving the Blitz. Iconic Images of the bombed cities during WWII and their instrumentalization for current conflicts in Britain and Germany." Paper presented at Constructions of Conflict: Transmitting Memories and the Past in European Historiography, Literature and Media, MEICAM, University of Swansea, 10–12 September 2007.

Slager, Kees. De Ramp. Een reconstructie van de watersnood van 1953. Amsterdam and Antwerp: Uitgeverij Atlas, 2003.

Sontag, Susan. "The Imagination of Disaster." Commentary October 1965: 42–48.

van Vree, Frank. In de Schaduw van Auschwitz: herinneringen, beelden, geschiedenis. Groningen: Historische Uitgeverij, 1995.

Internet:

Assmann, Aleida. "Impact and Resonance – A culturalist approach to the emotional deep structure of memory." 2010. http://www.liv.ac.uk/soclas/conferences/Theorizing/Kurzfassungok2.pdf (accessed 7 January 2012).

Illies, Florian. "Die Macht der Bilder." *Die Zeit*, 17 March 2011. http://www.zeit.de/2011/12/Iconic-Turn-Bildermacht (accessed 11 November 2011).

Pfister, Christian. " 'The Monster Swallows You!' Disaster Memory and Risk Culture in Western Europe, 1500–2000." Rachel Carlson Center Perspectives, January 2011. http://www.carsoncenter.uni-muenchen.de/publications/perspectives_mainpage/old_perspectives/rcc_issue_4web.pdf (accessed 11 January 2012).

Thijs, Krijn. "Niederlande – Schwarz, Weiß, Grau. Zeithistorische Debatten seit 2000." Version: 1.0, in: Docupedia-Zeitgeschichte, 3. 6. 2011. https://docupedia.de/[zg]/Niederlande_-_Schwarz_Weiss_Grau?oldid=79478 (accessed 7 January 2012).

Authors

Daniela Agostinho holds a PhD in Culture Studies from the Catholic University of Portugal, where she is an Invited Assistant Professor. She is founding editor of *Diffractions – Graduate Journal for the Study of Culture* and co-editor of *Panic and Mourning. The Cultural Work of Trauma* (Berlin: Walter de Gruyter, 2012). Her main areas of interest are Cultural Theory, Visual Culture, Film and Memory Studies.

Elsa Alves, Ph.D. fellow in Arts and Cultural Studies at the University of Copenhagen and researcher affiliated to COPE (Copenhagen Centre for Disaster Research) and CECC (Research Centre for Communication and Culture). She was the research assistant with a grant from FCT (Portuguese Foundation for Science and Technology) of the joint project "Critique of singularity: the catastrophic event and the rhetoric of representation", between the Catholic University of Portugal, the FU Berlin and the University of Kyoto (2010–13).

Claudia Benthien, Professor of Germanic Literatures and Cultural Theory at the University of Hamburg. Her work focusses on historical anthropology, cultural theory, gender studies, intellectual history, intermedial studies and visual culture. She is currently working on a research project on "The Literariness of Media Art".

Gabriele Brandstetter, Professor of Theatre and Dance Studies at Freie Universität Berlin. Her research focus is on: History and aesthetics of dance from the eighteenth century until today, theatre and dance of the avant-garde; contemporary theatre and dance, performance, theatricality and gender differences; concepts of body, movement and image. Selected publications: *Tanz-Lektüren. Körperbilder und Raumfiguren der Avantgarde* (1995); *ReMembering the Body. Körper-Bilder in Bewegung* (2000); *Bild-Sprung. TanzTheaterBewegung im Wechsel der Medien* (2005); *Schwarm(E)Motion. Bewegung zwischen Affekt und Masse* (2007); *Tanz als Anthropologie* (2007); *Notationen und choreographisches Denken* (2010); Dance [and] Theory (2013).

Isabel Capeloa Gil, Professor of Cultural Theory at the Catholic University of Portugal and Senior Researcher at the Centre for Communication and Culture (CECC); director of The Lisbon Consortium. Her main research areas include intermedia studies, gender studies, representations of war and conflict. Her most recent publications include *Mythographies. Figurations of Antigone, Cas-*

sandra and Medea in German Twentieth Century Drama (2007), *Landscapes of Memory. Envisaging the Past/Remembering the Future* (with Richard Trewinnard, 2004); *The Colour of Difference: On German Contemporary Culture* (with Mónica Dias, 2005), *Fleeting, Floating, Flowing: Water Writing and Modernity* (Würzburg, 2008) and *Kulturbau* (with Peter Hannenberg, Würzburg, 2009), *Visual Literacy. On the Disquiet of Images* (Lisbon, 2011). Editor of the international peer-reviewed journal *Comunicação e Cultura* (*Communication and Culture*) and member of the editorial boards of several European journals. She has been visiting professor at several institutions in Europe and the U.S., Fulbright Scholar at Western Michigan University in 2001 and is Honorary Fellow at the Institute of Germanic and Romance Studies (London).

Carmen Diego Gonçalves, PhD, Associate researcher at GEG-IGOT, Lisbon, and collaborator at the European Centre for Urban Risk. Was a post-doctoral researcher, with a grant from FCT at Center for Social Studies, Coimbra, with research focusing on dimensions of vulnerability and resilience, exploring the composite nature of the concept of natural disaster. Her fields of interest and research are: Theory of thinking styles and social representations; conceptions and perceptions of risk; risk societies; disasters and catastrophes; vulnerability and resilience; science, technology and society; knowledge and scientific practices; science communication and vulgarization; controversies and public participation; communication strategies; qualitative and quantitative methods of data collection and analysis.

Diana Gonçalves, PhD in Culture Studies, jointly awarded by the Catholic University of Portugal and the Justus-Liebig University, Giessen. Researcher at the Research Center for Communication and Culture (CECC), before that, member of the research project "Critique of Singularity: Catastrophic Events and the Rhetoric of Representation". Her main areas of interest are Culture Studies, American Studies, Translation Studies and Media Studies.

Isak Winkel Holm, teaches and researches at the University of Copenhagen, Department of Arts and Cultural Studies. His field of research implies German, French and Scandinavian literature and philosophy after 1750. His research deals with the relationship between literature and justice or, more generally, between aesthetics and politics. Hence, his research themes fall in two halves: on the one hand, Literary Theory, Cultural Theory, Aesthetics, Theory of Metaphor and Theory of Tragedy; on the other hand, Political Theory, Social Philosophy, Theory of Justice and Disaster Research. Currently, the two halves meet in two book projects.

Fernando Ilharco, Prof. at the Catholic University of Portugal and researcher at the Research Centre for Communication and Culture (CECC) at the School of Human Sciences, Lisbon. His PhD is from the London School of Economics and Political Science (LSE), London (2002). Since then he has published regularly at scientific publishers such as Oxford University Press, Wiley, Macmillan, Springer, Elsevier and Universidade Católica Editora.

Toshio Kawai, Ph.D., professor at the Kokoro Research Center, Kyoto University. He is also a Jungian analyst. He was educated in clinical psychology at Kyoto University and in philosophical psychology at Zurich University where he received a Ph.D. in 1987. He obtained his diploma from the C.G. Jung Institute Zurich in 1990. He has published articles and book chapters in English, German, and Japanese. His research field is clinical psychology, especially the cultural and historical background of psychotherapy and how consciousness is reflected in psychotherapy today.

Paulo de Medeiros, Professor of Modern and Contemporary World Literatures, teaches on the English and Comparative Literary Studies program. He was Associate Professor at Bryant College (USA) and Professor at Utrecht University (Netherlands) before moving to Warwick. In 2011–2012 he was Keeley Fellow at Wadham College, Oxford and is currently President of the American Portuguese Studies Association. Current projects include a study on Postimperial Europe. His research interests include World Literatures, Modernism and Postcolonial Studies, and Lusophone Literatures. His qualifications are BA; MA (Massachusetts at Boston), MA: PhD (Massachusetts at Amherst).

José Manuel Mendes, Professor of Sociology at the School of Economics of the University of Coimbra, Portugal. He is also a senior researcher at the Centre for Social Studies of the University of Coimbra, where he has been working in the fields of inequalities, social mobility, social movements and collective action and, more recently, on the themes of risk, social vulnerability, trauma and victims' associations. He is coordinator of the Risk Observatory (OSIRIS) of the Centre for Social Studies. Among his publications is the edited book (with Pedro Araujo), *Os lugares (im)possíveis da cidadania: Estado e risco num mundo globalizado* (The (im)possible places of citizenship: risk and state in a globalized world). Coimbra: Almedina, 2013.

Johana E. Prawitasari-Hadiyono, acquired her Ph.D at University of Arizona, USA. She has been Professor of Clinical Psychology at Gadjah Mada University,

Yogyakarta and is now working in the same field at UKRIDA (Universitas Kristen Krida Wacana) in Jakarta.

Joan Ramon Resina, Professor of Iberian and Latin American Cultures and Comparative Literature, Director Iberian Studies Program, Chair of Graduate Studies, Iberian and Latin American Cultures. Most recently author of *Del Hispanismo a los Estudios Ibéricos. Una propuesta federativa para el ámbito cultural.* Madrid: Biblioteca Nueva, 2009, *Barcelona's Vocation of Modernity: Rise and Decline of an Urban Image* (Stanford UP, 2008). He has also published extensively in specialized journals, such as *PMLA, MLN, New Literary History,* and *Modern Language Quarterly.* From 1999 to 2005 Editor of *Diacritics.* Has held teaching positions at Cornell University, the State University of New York at Stony Brook, and Northwestern University, as well as visiting appointments at foreign universities, and received awards such as the Alexander von Humboldt and the Fullbright fellowship.

Shoko Suzuki, educated in Japan and Germany, receiving her Dr. Phil. from Sophia University in Tokyo 1987. She is Vice Dean, Professor at Graduate School of Education Kyoto University in Japan, Member of The Science Council of Japan, Academic Leader of Inter-Graduate School Program for Sustainable Development and Survivable Societies in Kyoto University; Research activities University of Cologne 1982–1989 and Guest Professorship at Freie Universität Berlin 2009/10. Publications: "Takt in Modern Education", Münster, New York 2010; (with Christoph Wulf): Auf dem Weg des Lebens. West-östliche Meditationen. Berlin 2013.

Eduardo Cintra Torres, Assistant Professor, Faculty of Human Sciences, Catholic University of Portugal, and ISCTE-IUL. Researcher at CECC. PhD in Sociology. Author of 14 books, the latest being *Multidão e Televisão: Representações Contemporâneas da Efervescência Colectiva* [The Crowd and Television: Contemporary Representations of Collective Effervescence], Lisbon, UCE, 2013. Author of book chapters and articles in Portuguese, English and French, published in Portugal, France, Brazil and Canada, author of pedagogic materials for the Ministry of Education and of television and radio programmes for Portuguese and Macao televisions. Journalist since 1983. TV and media critic in daily *Público* (1996–2011) and *Correio da Manhã* (2011–) and advertising critic in daily *Jornal de Negócios* since 2003.

Inês Espada Vieira, assistant professor at the Faculty of Human Sciences and a researcher at the Research Centre for Communication and Culture, Catholic

University of Portugal. She holds a PhD in Culture Studies from the same university (European PhD). Currently, she is the Coordinator of the B.A. in Applied Foreign Languages and member of the Board of the Faculty of Human Sciences. She has published articles on Portuguese and Spanish literature and culture and has co-edited, among others, *Intellectual Topographies and the Making of Citizenship* (2011). She is the author of *Intellectuals, Modernity and Memory* (2012)

Christoph Wulf, Professor of Anthropology and Education and member of the Interdisciplinary Centre for Historical Anthropology and the Graduate School "InterArts" at Freie Universität Berlin. Vice President of the German Commission for UNESCO. Recent Publications: *Anthropology – A Continental Perspective*. Chicago 2013: University of Chicago Press; *Das Rätsel des Humanen*. Munich 2013: Wilhelm Fink. *Major research areas:* historical and cultural anthropology, educational anthropology, rituals, gestures, emotions, intercultural communication, mimesis, aesthetics.

Manfred Zaumseil, 1979–2008 Professor for Clinical and Community Psychology at Free University of Berlin. After retirement affiliated to "Internationale Akademie für Pädagogik, Psychologie und Ökonomie gGmbH". Current fields of work: culture & mental health, psychosocial consequences of disasters. Recent publication: Zaumseil, M., S. Schwarz, M. Vacano, G.B. Sullivan, J.E. Prawitasari-Hadiyono (eds.), Cultural psychology of coping with disasters. The case of an earthquake in Java, Indonesia; New York 2014.

Index of Names

www.ingramcontent.com/pod-product-compliance
Lightning Source LLC
Chambersburg PA
CBHW050645270326
41927CB00012B/2877